BURLEIGH DODDS SCIENCE: INSTANT INSIGHTS

NUMBER 99

Improving the welfare of growing and finishing pigs

bd burleigh dodds
SCIENCE PUBLISHING

Published by Burleigh Dodds Science Publishing Limited
82 High Street, Sawston, Cambridge CB22 3HJ, UK
www.bdspublishing.com

Burleigh Dodds Science Publishing, 1518 Walnut Street, Suite 900, Philadelphia, PA 19102-3406, USA

First published 2024 by Burleigh Dodds Science Publishing Limited
© Burleigh Dodds Science Publishing, 2024, except the following: the contribution of Luigi Faucitano in Chapter 3 is © Her Majesty the Queen in Right of Canada. All rights reserved.

British Library Cataloguing in Publication Data
A catalogue record for this book is available from the British Library

ISBN 978-1-80146-671-4 (Print)
ISBN 978-1-80146-672-1 (ePub)

DOI: 10.19103/9781801466721

Typeset by Deanta Global Publishing Services, Dublin, Ireland

Contents

Series list

Acknowledgements

Chapters in this Instant Insight are taken from the following sources:

Chapter 1 Optimising pig welfare in the growing and finishing stage
Chapter taken from: Edwards, S. (ed.), Understanding the behaviour and improving the welfare of pigs, Burleigh Dodds Science Publishing, Cambridge, UK, 2021, (ISBN: 978 1 78676 443 0; www.bdspublishing.com)

Chapter 2 Welfare of pigs during finishing
Chapter taken from: Wiseman, J. (ed.), Achieving sustainable production of pig meat Volume 3: Animal health and welfare, Burleigh Dodds Science Publishing, Cambridge, UK, 2018, (ISBN: 978 1 78676 096 8; www.bdspublishing.com)

Chapter 3 Optimising pig welfare during transport, lairage and slaughter
Chapter taken from: Edwards, S. (ed.), Understanding the behaviour and improving the welfare of pigs, Burleigh Dodds Science Publishing, Cambridge, UK, 2021, (ISBN: 978 1 78676 443 0; www.bdspublishing.com)

Chapter 4 Optimising the health of finisher pigs
Chapter taken from: Maes, D. and Segalés, J. (ed.), Optimising pig herd health and production, Burleigh Dodds Science Publishing, Cambridge, UK, 2023, (ISBN: 978 1 78676 883 4; www.bdspublishing.com)

Chapter 1

Optimizing pig welfare in the growing and finishing stage

Arlene Garcia and John J. McGlone, Texas Tech University, USA

1 Introduction

The US pig inventory has increased by 15% in the past 3 years, with a pig population greater than 71 million (NASS, 2017). Global pork production has increased fourfold over the past 50 years, driven by population growth and dietary transition toward more animal protein per capita (Lassaletta et al., 2014, 2019; Bai et al., 2018a). The US markets over 100 million pigs per year (about 10% of the world production).

In 2018, China had about half of the world's numbers of pigs. In that year, China and other Asian countries broke with African Swine Fever (ASF). Prior to this outbreak, pork was the meat consumed in the highest amount in the world, mainly due to a high level of consumption in Asia. Estimates are that at least half of the pigs in China have been culled due to ASF. This means that the world's pig population decreased in a short period by over 25%. ASF has also broken out in Southeast Asia, Russia, Poland and Belgium. It is thought to be carried among farms by feral or wild pigs in Europe. If this is so, ASF may spread to other European countries. When ASF breaks in a country, that country is not allowed to export pig meat. This changes the economics of the infected country. The concern is that when pig prices are low, or when large numbers of pigs must be euthanized, that animal welfare may suffer. Currently,

http://dx.doi.org/10.19103/AS.2020.0081.06

in 2020 a pandemic that began in China (in December 2019) has spread around the world, and many countries and regions have 'stay-at-home' orders. How this pandemic will impact the world's swine business is yet to be determined. If farm and slaughter plant workers do not work, then the pork supply (and all food) will decrease. Additionally, the inability of employees to go to work negatively impacts the welfare of pigs on farms. The extent of a pandemic on pig welfare will most likely be studied later.

ASF virus (ASFV) has a complex transmission cycle involving African wild suid species, soft ticks, and domestic pigs (Sánchez-Vizcaíno et al., 2019). ASFV is transmitted, most often directly via contact between sick and healthy animals, including domestic pigs and wild boars (Sánchez-Vizcaíno et al., 2019). As the disease progresses, animals may stop eating and become increasingly moribund, presenting bloody diarrhea, vomiting and possibly abortion (Dixon et al., 2019). Pathological changes associated with vasculitis include skin erythema, pulmonary edema, hyperemic splenomegaly, hemorrhagic lymphadenitis and petechial hemorrhages in the kidneys, lungs and urinary bladder (Dixon et al., 2019). Dixon et al. (2019) reported that infection is associated with very high levels of virus in the blood (up to 109 TCID50/ml) and tissues, particularly the spleen and lymph nodes. ASF is a hemorrhagic disease that causes illness quickly with a very high mortality rate (Dixon et al., 2019). The usual practice is to euthanize all the pigs on the infected farms and on all farms near the infected farm. ASF causes two main welfare problems. First, infected animals suffer a horrible death. Second, euthanizing so many pigs is a challenge, even in countries with sophisticated animal welfare rules.

As a result of the Asian ASF outbreak, pig meat is being exported from many exporting countries to China. About 25% of the pork in the USA is exported. In 2019, the USA and China were in a trade war. This results in little US pork going to China. This lowers the pig price in the USA since there is an oversupply without the Chinese market. The rest of the world has high pig prices due to the increased demand and because China must import more pork than in the past. The ASF outbreak will increase the world's pork prices until the disease is eradicated, yet there is no vaccine for ASF. To meet this demand, new farms will be built using current intensive technologies. Smaller farms were hit harder with ASF than modern-style, intensive farms. The regrowth in the Chinese pig industry, which is under way now, will be primarily large, intensive pig units. ASF accelerated the move away from small, low-input farms to larger, intensive pig units. Small, backyard farms and large intensive farms have very different welfare concerns.

The majority of pigs in pork-exporting countries are raised in intensive production systems. The trend is for the world industry to continue to restructure with more intensive farms and fewer small farms. ASF accelerates this trend greatly in infected and pork-exporting countries (USA, Brazil, Germany, Spain, Denmark, The Netherlands and other EU countries).

In intensive farms in the USA, pigs are weaned from the sow around 3 weeks of age (while European directives do not allow weaning until 28 days of age) and moved to a nursery (or a wean-to-finish barn, bypassing the nursery) where they will stay for 6-8 weeks, and then to a grow-finishing unit where they will spend 16-17 weeks until they reach market weight (National Pork Board, NPB, 2019). When pigs are moved to the finishing barn, they are termed grow-finishing pigs and are fed there until they reach market weight (at about 280 lbs/127 kg in the USA); then they are termed market pigs.

Intensive production systems are where animals are kept and raised within fewer production units, with a large number of animals within these units (Fraser, 2005). Such intensification can have negative effects if not managed properly, including increased animal aggression, competition over resources, disease, injuries and death, and over-all poor animal welfare. However, intensive production systems can be well controlled (Clark et al., 2019) and have fewer animal welfare issues if they are properly managed.

Societal concerns about the well-being of livestock animals are a driving factor for the swine industry to adopt practices and systems that are pig welfare friendly. As the growing demand for animal protein grows, and hence a shortage of labor, there is more need for automated systems, but caretakers play a vital role in ensuring these systems work properly. Caretakers are responsible for the day-to-day care of pigs and identifying problems within the barn. Currently, we ultimately rely on human observation for the welfare of pigs.

Real-time monitoring of pigs in large-scale settings has become crucial (Nasirahmadi et al., 2017), as the direct observation of every animal is intermittent and impractical (Millman, 2007; Weary et al., 2009). New smart barns may allow the automation of animal welfare checks, among other benefits.

The main objectives of this chapter are to define animal welfare and to identify where intensive production systems may put finishing pigs at risk of poor animal welfare.

The first challenge in assessing animal welfare is to define the term. Its definition varies, which adds to the complexity of this issue. The OIE (Terrestrial Code) animal welfare code defines animal welfare as 'the physical and mental state of an animal in relation to the conditions in which it lives and dies' (OIE, 2010). The OIE further guides its definition of animal welfare through the 'guiding principles', known as the 'Five Freedoms'. The Five Freedoms were proposed in 1965 and describe society's expectations for the conditions animals should experience under human control (OIE, 2010). The Five Freedoms encompass the following:

- Freedom from hunger, malnutrition and thirst;
- Freedom from fear and distress;

- Freedom from heat stress and physical discomfort;
- Freedom from pain, injury and disease; and
- Freedom to express normal patterns of behavior.

However, animal welfare can be defined in a number of ways, but, whatever the definition is, some authors believe it must include three elements (Velarde et al., 2015): the biological function of the animal (or ability to cope with its environment), its emotional state, and its ability to show normal patterns of behavior (Manteca et al., 2009).

Animal welfare is perceived as a 'public good' (Miele and Evans, 2010) and is a necessary element of sustainable animal production (McGlone, 2001; Broom, 2010). Sustainable meat production is 'economically sound, economically viable, socially just, and humane' (Appleby, 2004). Further, animal welfare is a matter of ethics and an essential tool to gain and maintain markets, and husbandry that benefits sustainability and maximizes the welfare of animals, but avoids its potential impairment (Velarde et al., 2015).

2 Pig behavioral issues that impact pig welfare

Welfare issues may arise in pig production when there is a mismatch between a pig's instincts or behavior and its environment (Kittawornrat and Zimmerman, 2011). Behavioral impulses may be expressed inappropriately when instinctual behavior is thwarted (Kittawornrat and Zimmerman, 2011). In the wild, pigs are active at night and spend the majority of their awake time in foraging-related activities, including rooting, grazing and exploring with their snout (Caley, 1997), while outdoor domestic pigs are active during the day (Stolba and Wood-Gush, 1984). Pigs in commercial settings are active in the day and continue to express exploratory behaviors (Kittawornrat and Zimmerman, 2011).

2.1 Oral-nasal-facial/oral behaviors

Oral-nasal-facial (ONF) behaviors are behaviors associated with the snout, nose and face of the pig. Pigs in both indoor and outdoor systems express ONF behaviors (Dailey and McGlone, 1997). Some people call them oral-nasal behaviors, but the detailed work by Dailey and McGlone (1997) showed that the face is also used to interact with substrates. Pigs kept outdoors have a much higher rate of ONF behaviors than pigs kept indoors. ONF behaviors include chew/bite grass, chew/bite fence/bars, chew rocks/soil and rooting the ground or trough. Indoor pigs have less material to manipulate than pigs on a pasture or wooded area. This lack of material to manipulate may be frustrating to indoor pigs, which can lead to what we call abnormal behaviors, which may not be abnormal if these are appropriate behaviors for the environment. Certainly, pork producers would like to minimize damaging ONF behaviors. Providing rooting materials may reduce unwanted damage toward animals or penning and direct these behaviors to rootable substrates. Dailey and McGlone (1997) concluded that ONF-stereotyped behaviors may be natural pre- and post-feeding appetitive and consummatory chewing. Rooting activities in sequences form by the available substrates.

2.2 Tail biting

Tail biting is often an indication that welfare is low in the pen as a whole (Fritschen and Hogg, 1996). There are a variety of management practices that can reduce tail biting outbreaks, such as the provision of enrichment (Beattie et al., 2001), removal of the tail biter or removal of the victim (Zonderland et al., 2008), but some farms at certain times cannot control tail biting. Further research in this area is needed to find one or more solutions.

2.3 Belly nosing

Belly nosing is the act of rhythmic up-and-down movement of one piglet rubbing the belly of another with its snout (Fraser, 1978). This behavior can lead to skin lesions and ulcerations on the belly and flank of the victim (Straw and Bartlett, 2001).

When piglets are raised with the sow, piglets orient toward the udder and massage the sow's underline to induce milk letdown as an example of ONF behaviors expressed within a few minutes of birth (Frei et al., 2018). Piglets raised without a sow develop belly nosing behaviors (Weary et al., 1999; Rzezniczek et al., 2015), and the prevalence increases with decreasing weaning age (Metz and Gonyou, 1990; Worobec et al., 1999). Studies suggest that belly nosing is not an indicator of stress (Gardner et al., 2001a), not influenced by diet or diet quality or the presence of milk in the diet (Gardner et al., 2001b), but is more closely related to social interactions than with eating or drinking (Li and Gonyou, 2002); thus it is a type of redirected ONF sucking behavior (Fraser, 1978). Further, piglets reared with the sow do not show belly nosing behaviors (Rzezniczek et al., 2015), which may indicate the behavioral needs of piglets for a specific kind of ONF behavior are not met in artificial rearing systems (Frei et al., 2018).

Commercial environments are often barren and prevent pigs from displaying natural behaviors, such as rooting; thus an environment with poor stimuli can contribute to the development of what are called abnormal behaviors (Cooper et al., 2001; Bolhuis et al., 2005), but may be perfectly normal given the pig's development and inappropriate environment. The inability to express species-specific behavior in foraging in a barren environment can increase aggression by enhancing competition, leading to more negative agonistic behavior (Docking et al., 2008), as well as to frustration, a stress factor itself (Casal-Plana et al., 2017).

2.4 Aggression

During a pig's life cycle, there may be several aggressive interactions with other pigs, which can result in injuries and stress (Hemsworth et al., 2015) and detrimentally affect productivity (McGlone et al., 1980; Verdon and Rault, 2018), at least temporarily. Aggression can be observed during the mixing of pigs, as a social hierarchy is established within the group (McGlone, 1986) and typically declines after the hierarchy is established (Langbein and Puppe, 2004).

Mixing of unfamiliar pigs leads to aggression and can negatively impact physiology and welfare (Arey and Edwards, 1998; Einarsson et al., 1996; Knox et al., 2014; Rault et al., 2014; Verdon et al., 2015) due to fighting when pigs form social hierarchies and during feeding (Arey and Edwards, 1998; Li et al., 2012). These aggressive encounters can result in injury, lameness and increased cortisol concentrations (Anil et al., 2006; Arey and Edwards, 1998; Li et al., 2012) and can affect weight gain (Mormède et al., 2007).

When unfamiliar pigs are mixed in intensive units, they fight. At a minimum, intensive pigs are mixed at weaning and when they are transported to slaughter. Some pigs are also mixed between growing and finishing phases and in late finishing when regrouping smaller finishing pigs, as heavier pigs in the barn are marketed.

When outdoor pigs are weaned and mixed, they do not fight, while indoor pigs do fight when weaned and mixed (Sarignac et al., 1997). Outdoor pigs mingle among different litters during lactation. Social behaviors are practiced, and this behavioral practice is thought to reduce fighting later because piglets learn appropriate social skills. Pigs kept with a single sow in a farrowing crate may lack the social skills needed to prevent fighting. In this thinking, the indoor system causes pig aggression due to a lack of social skills in developing piglets, which is directly caused by the intensive production system.

A commonly overlooked factor that leads to aggression in the herd is the increase in pig weights over time. Increased pig weight and weight gain can lead to increased motivation to eat and chew and can cause stocking density issues toward the end of the finishing phase, leading to competition over resources. Market weight has linearly increased by 5.8 kg every 10 years for the past 4 decades, driven by the dilution of fixed cost over more weight per pig, improved genetics and nutrition resulting in more efficient pigs (Brumm, 2012). Higher pig weights are essentially driven by packer demands for carcasses that can provide more kg of lean meat to improve packing plant throughput. Producers consistently respond to market demands, and one of the market demands has been for greater market weights that produce carcasses with greater kg of lean meat per carcass.

Although pig market weights vary depending on regions and cultural backgrounds, as of the beginning of 2017 in the USA, the average market weight was reported to be 129 kg, but varied between 124 kg and 130 kg in 2016 (NASS, 2017). The design and management of commercial group housing are complex and with wide disparity (Grandin and Whiting, 2018). Proper facility management can decrease pig aggression and lead to a positive impact on pig welfare (Grandin and Whiting, 2018). Such management strategies should focus on controlling aggression and protecting vulnerable pigs from aggression. This may include minimizing changes in the social structure (i.e. mixing), modifying pen design to allow escape from aggressors (McGlone and Curtis, 1985), incorporating distractions at regrouping, reducing stocking density and facilitating access to resources such as feed and water (Marchant-Forde and Marchant-Forde, 2005).

3 Production systems

Pigs are typically weaned at 21 days and moved to a nursery. Confinement systems for pigs are generally confronted by skepticism and concern from the public (Sørensen and Schrader, 2019) because of the negative impact they can have on animal welfare. However, any poorly managed facility can negatively affect animal welfare.

Most confined systems follow the all-in/all-out (AIAO) management system. The AIAO strategy has many advantages for pig production, if it is adhered to, which many times does not happen. The benefits of AIAO include improved biosecurity/minimized disease transmission, health, and growth performance (Scheidt et al., 1995; Owsley et al., 2013). Pigs in the AIAO system are matched by age and move forward through the production stages in the same groups without re-mixing pigs of different ages (Owsley et al., 2013). The rooms these pigs occupied are, therefore, left empty and are cleaned and disinfected for the next group of pigs (Diana et al., 2019). Adhering to an AIAO management strategy is reported to have major constraints. These constrains include the lack of facilities to house slow-growing pigs/sick pigs (Diana et al., 2019), leading to the practice of sorting pigs by body weight into the next production stage (Calderón Díaz et al., 2017) to achieve uniformity in slaughter weights (Brumm et al., 2004). Slow-growing/sick pigs disrupt the normal flow by being delayed for several weeks from moving to the next stage and are then re-mixed with pigs of their own size, but not age (Diana et al., 2017). This practice is reported to increase disease transmission between different age groups, can have an adverse effect on pig performance, can cause pigs to be 10 kg lighter at slaughter and have increased risk of diseases such as pleurisy and pericarditis compared to the pigs in the 'normal' flow (Calderón Díaz et al., 2017). The stress associated with re-mixing these slow-growing/sick pigs leads to the development of damaging behaviors such as fighting wounds and ear and tail biting (Schrøder-Petersen and Simonsen, 2001; Martínez-Miró et al., 2016), reflecting poor welfare (Schrøder-Petersen and Simonsen, 2001).

3.1 Grow-finish pigs in alternative housing

Indoor, mechanically ventilated systems are generally chosen over other systems such as hoop and pasture systems due to their lifetime use of 15-20 years or longer (Kliebenstein et al., 2003), the ability to have environmental control (ventilation, misters, heating system), increased biosecurity and other factors that can be controlled. Under year-round pork production, confinement systems generally have superior pig performance and lower variable costs (such as feed costs) compared to hoop or pasture systems (Larson et al., 2002, 2003).

Due to the increased concern about animal welfare, there is an interest in alternative production systems, such as straw-based systems, straw yards and outdoor production systems (Guy et al., 2002). In a study by Guy et al. (2002), genotype and finishing systems were compared to identify an interaction. The welfare of progeny from outdoor/semiintensive-type sows and indoor sows was evaluated, in addition to pig welfare in straw yards (considered a less-intensive system), outdoor paddocks and conventional, fully slatted pens. They found that the interaction between the housing system and genotype for health parameters was not significant. However, Guy et al. (2002) reported that outdoor paddocks and straw yards enhanced pig health and welfare compared to conventional, fully slatted pens, shown by the levels of stomach ulcers, lung damage, bursitis and body damage.

Most pigs in developed countries are finished indoors in pens with concrete slats and mechanical ventilation. The intensity of indoor housing has been increasing in recent years, with the introduction of what is called filter barns (Reicks, 2009). In these barns, the incoming air is filtered to prevent microbes from entering the building. This means the barns must be sealed to any outside air, and mechanical ventilation is the only way air is moved in and out. We have not seen studies that attempt to compare the welfare of pigs in a filter barn with a conventional barn. However, today the non-filtered grow-finish barn with concrete slats and mechanical ventilation is the industry standard. With the advent of highly infectious diseases like ASF or porcine reproductive and respiratory syndrome (PRRS), filter barns are likely to be more common in the future.

An alternative system must be different from the standard indoor building we have today. While people suggest that alternative systems are more humane, the main reasons to use an alternative barn are often the lower cost to build or use the structure. Three common alternative systems (from most to least intensive) are bedded poultry barns, hoop structure and free range with a shelter. While there are many others, these three have at least some published data.

Twenty years ago, Jeff Hill (2000) reviewed deep-bedded finishing barns and hoops. He identified that there should be considerations of animal performance, environmental issues, marketing options, initial facility investments, and animal welfare issues. This section will touch on these issues but focus on animal welfare and on more recent literature. Most lay people observing a well-managed bedded or outdoor system would prefer that system over a more intensive system. However, alternate systems sometimes have serious welfare issues if they are not managed properly. We must also consider other society issues when a system is considered (McGlone, 2001). At least we need to consider animal welfare, food safety and environmental impacts of any outdoor or indoor-bedded facility.

One society issue has trumped all others in the recent disfavor of alternative systems: food safety. In the past (15-20 years ago), we could keep *Salmonella* out of indoor pigs, but we could not keep it out of the outdoor system, especially the wallows (Callaway et al., 2005). More recently, porcine epidemic diarrhea virus (PEDv) and ASF have moved among outdoor herds freely and unpredictably by birds, wild pigs and rodents. These infectious diseases (and probably more) are deal-killers for outdoor systems in the USA and parts of Europe and Asia. Still, outdoor and bedded facilities are favored at first glance by consumers.

3.2 Flooring

Slatted concrete floors are the most commonly used type of floor in intensive units due to their effective drainage of manure (Devillers et al., 2019). However, slatted floors can have negative effects (such as lameness), but limited studies have been conducted to identify the effect of slatted concrete flooring design on pig performance and welfare (Devillers et al., 2019).

Slat design has been studied for manure drainage and ammonia emissions, with slat width based on the efficiency of manure drainage (Aarnink et al., 1996, 1997; Vermeij et al., 2009; Ye et al., 2009). In North America, slats usually measure 5 inches and gaps measure 1 inch in width, with a permeability of 17% for a fully slatted floor (Devillers et al., 2019).

Rahse and Hoy (2007) found a tendency for more severe claw lesions in fattening pigs kept on slatted floors with a gap wider than 20 mm (0.78 inches), whereas Falke et al. (2018) found no effect on the slat width on leg and foot lesions between 153 mm (6.02 inches) and 190 mm (7.5 inches) slats, both with gaps of 15 mm (0.59 inches). The relatively high prevalence of toe lesions may be linked to pigs being predominantly housed on concrete floors (Cameron,

2012; Penny et al., 1965; Schulenburg et al., 1986). When pigs are housed on slatted floors, an increased risk of sole and heel erosions has been reported and these lesions' prevalence is higher compared to pigs housed on solid floors with straw bedding (Jørgensen, 2003). Gentry et al. (2002a) examined feet and lungs (and meat quality) of indoor pigs raised on concrete slats with pigs raised on bedding. Pigs on bedding had fewer overall foot lesions, but the percentage of severe foot lesions was three times higher on bedding than for pigs on concrete slats. When pigs on bedding get a foot lesion, the lesion is more likely to develop into a severe foot lesion.

Webb (1984) reported that pigs walk mainly on their outer digits and also use their inner toe to some extent; however, the compressive strength of the hoof wall does not increase in heavier pigs. The peak pressure on the foot is almost independent of weight, and the total contact area varies almost linearly, which gives weight peak hoof pressures and hoof compressive strengths a safety factor for the foot (Webb, 1984). Gap recommendations on slat and gap width vary based on the size of the animal, mainly due to the size of the foot in proportion to the body weight and the void percentage of the slatted floor as a function of live weight (Jensen et al., 1997; Broom et al., 2005). Matching the pig foot size with the slats is important so that the manure flows, and the pigs do not get their feet stuck in the slats. Anatomical measurement of claw widths in sows is 25-33 mm (0.98-1.30 inches), and toes measure 45-51 mm (1.77-2 inches), on average (Broom et al., 2005; Sasaki et al., 2015). A study by Tubbs (1988) gave a recommendation of 13-16 mm (0.51-0.63 inches) rather than 19-25 mm (0.75-0.98 inches) to minimize injuries to toes and dew claws, which would then be more consistent with having slat width measure at least the sole length or twice the claw width (an average of more than 60 mm (2.40 inches); Devillers et al., 2019).

3.3 Bedded indoor and outdoor systems

A bedded indoor system is a way for pork producers to have pigs on bedding with a relatively low-cost building. Usually, these buildings are a re-purposed building that is converted for finishing pigs. These buildings became known when, in Missouri and surrounding states, the US packers needed more pigs and the finishing spaces were not enough in numbers. Figure 1 shows an example of a converted Turkey building in a region that has a large amount of fescue hull bedding from farms that produce fescue for seed. In the northern states, they use corn stalks, wheat, or barley straw for bedding. Because shipping bedding is expensive, producers use bedding that is available locally.

An important topic to consider with bedded facilities is the management of large amounts of wet or soiled bedding. This chapter does not review this issue, but it must be considered in practice.

Figure 1 Wide view (a) and close view (b) of finishing pigs near the end of the grow-finish cycle. Pigs are on fescue hull bedding, which is common in Missouri. The barn started with 2000 pigs, bedded, with natural ventilation (curtains go up and down). This barn was used to grow turkeys and re-purposed into a 'natural' finishing facility with bedding rather than concrete slats.

Two features of most bedded indoor facilities are the flooring (concrete slats or bedding) and group size (usually more than 500 pigs per barn, often 1000-2000 pigs).

The performance, health and welfare of pigs in bedded buildings versus conventional indoor slatted-floor systems have previously been studied. Gentry and others (Gentry et al., 2002a) examined pig growth, pork composition, and pork quality for pigs finished indoors on slats, and indoors on deep bedding (Fig. 2), or outdoor housing (Fig. 3). Pigs born and raised outdoors in the summer had higher weight gain and heavier carcasses and redder meat color than pigs in a more conventional indoor slatted barn. However, loins in a simulated retail display from the outdoor system in the summer had more discoloration and browning 4 days after display compared with indoor pigs. In another study, during the Texas winter, no differences were found in pig performance or in most meat quality traits. In the next comparison, cohorts of pigs were finished in either a conventional indoor barn or a deep-bedded barn with 1500 pigs in a single pen. Pigs on bedding grew faster and had

Figure 2 A row of hoop structures for growing-finishing pigs. Each barn held about 500 pigs. A tractor with a feed wagon was used to re-fill feeders.

Figure 3 An outdoor pig nursery (a) in Texas. Pigs will need to be moved when they get bigger to maintain their health and welfare. A Brazilian low-cost, very extensive system (b). As the pigs grow they will need to be moved to a larger space with more ground cover to improve their health and maintain the land. An outdoor finishing pen (c) in Texas. These pigs have a concrete slat section under the waterer to prevent pigs from making a wallow (there is a wallow in those pens away from the waterer). And the grass is green and growing in this field due to the low pig density.

heavier carcasses than indoor-reared pigs. More directly related to welfare, the authors measured foot and lung lesions from pigs on bedding and slats. The conclusion about the foot lesions is that, overall, bedding reduces foot lesions; however, if any pig begins to develop foot lesions on bedding, it will be much worse than if the pigs were on slats. For lung lesions, pigs on bedding had more overall lung lesions but fewer severe lung lesions than pigs on slatted floors.

Gentry et al. (2002a) further explored the mechanisms behind the improvements observed for pigs in the outdoor system. In their study, pigs born to contemporary sows which farrowed either indoors or outdoors were moved either indoors or outdoors. Overall, the outdoor-born pigs had higher growth and heavier carcasses than indoor-born pigs, regardless of being finished indoors or outdoors.

The same group examined muscle fiber types for indoor and outdoor-born pigs (Gentry et al., 2004). Loin chops from outdoor-born pigs had more type I and type IIA muscle fibers and a lower number of type IIB/X fibers than indoor pigs. Thus, the outdoor farrowing system changes muscle development compared to indoor-reared pigs in Texas. A similar study in Switzerland by Bee et al. (2004) performed in the winter did not find the same results as Gentry et al. (2004), but they measured different traits. Meat from outdoor-reared pigs had more drip loss and had fewer fast glycolytic fibers and different fatty acid composition than indoor pigs. These two investigators' findings did not actually conflict because the positive meat quality results by Gentry were found mostly in the summer.

To consider if the results above between housing systems were due to exercise among pigs in large pens or outdoors, Gentry et al. (2002b) used very narrow and long pens with the feeder on one end and the water on the other end. These pigs spent more time walking than pigs in a conventional finishing pen. Pigs forced to exercise did not have many differences in pig performance or meat quality. These results (Gentry et al., 2002b) cannot be explained simply by pigs in alternative systems getting more exercise. Other, yet-to-be-determined, factors are responsible for the differences in pig growth meat quality.

Clearly, more is to be learned about housing system effects on muscle fat/lean composition, metabolism and muscle fiber types. The effects of rearing systems on muscle fibers and collagen were reviewed 15 years ago by Gondret et al. (2005). Additionally, we need to focus on pig factors and less on growth and meat quality. Food safety, pig health issues and environmental issues should be considered and can be challenging. For pig factors related to animal welfare, the bedded facilities have a positive view from casual observers and there is some evidence that pigs are more active and more social in larger groups with bedding or outdoors.

3.4 Hoop buildings

A hoop building is a special type of bedded facility (Fig. 3). The structure can be made of metal or different kinds of heavy plastic. The inside is bedded, although we have seen an inside with slats. In this section, we refer to bedded hoop buildings. Some research was done on this housing system 15–25 years ago. Little current research can be found in the literature.

In an attempt to directly assess pig welfare in hoops and slatted facilities, Lay et al. (2000) reported that compared to pigs on slats, pigs in hoops had lower cortisol, fewer aberrant behaviors, more play and fewer injuries. We speculate that the hoop system was well managed. If the bedding is wet, especially in the winter, we may find different effects. A report from Portugal by Araújo et al. (2016) found no meaningful differences in pig welfare for a special Portuguese breed in hoop buildings, but the study had an unclear control treatment group.

Honeyman (2005) summarized his work in a 2005 review. This author has conducted the most extensive work on hoop structures in the USA. Honeyman (2005) concluded that pork quality was only slightly different for pigs in hoops compared to conventional indoor, slatted-floor facilities. He reported that pigs in hoops tended to be fatter (as we found with pigs in bedded indoor facilities). This may indicate that pigs in bedded facilities need to be fed differently. Also, this finding may indicate that the bedding creates a warmer microenvironment, and with a similar feed intake, they would be fatter. It is interesting that Honeyman reports smaller carcass weight in hoop buildings, while in our bedded poultry facility, the carcass weights were heavier than in indoor slatted facilities. Part of this may be explained by differential effects depending on the season.

A common difference for pigs in hoops is that they are often sold in a 'natural' pork scheme of various types. As such, they are often not fed antibiotics. The comparison of indoor slated facilities with hoop or bedded or outdoor facilities needs to have similar antibiotic applications and a common cohort of pigs (same genetics and previous experience). For the most part, some data show similar performance and health issues in a variety of systems if the study has controlled for known factors (temperature, genetics, feed, health, time, etc.). The main issues in the USA are the significant food safety and animal health challenges with alternative systems – these were enough to limit the use of alternative systems except in small, specialized niche markets.

4 Group size

Marginal profit increases with the size of pig operations (Martin and Kruja, 2000); the swine industry is shifting toward larger production units. Thus, housing large numbers of similarly aged animals and housing pigs in large groups can reduce housing costs (Schmolke et al., 2003).

It has been suggested that as the group size increases, variation in performance also increases (English, 1988). Housing large groups of pigs has long been reported to have a negative impact that may lead to poor performance, increased variation in body weight (BW) and a higher incidence of behavioral vices (English, 1988). Group size can affect the performance of pigs, although welfare is not necessarily compromised when increasing the number of individuals within a group for a given space allowance per pig (Turner et al., 2003; Turner and Edwards, 2004). Group size and space allowance per pig are reported to often be confounded in many studies, making the interpretation of these two factors difficult (Averós et al., 2010).

Schmolke et al. (2003) found that average daily gain (ADG) in groups of 40 pigs was 12% lower than pigs in groups of 10 during the initial 2 weeks of their study. Spoolder et al. (1999) reported a 5% reduction in the growth rate in groups of 40-80 pigs up to 65 kg but not during the finishing period, compared with pigs in groups of 20. Wolter et al. (2001) conducted a wean-to-finish study where he found that growth rates were poorer in groups of 50 and 100 than in groups of 25 between the weights of 6 kg and 35 kg. However, Wolter et al. (2001) did not find differences in the final BW and overall growth rate among the group sizes. Schmolke et al. (2003) did not observe an increase in variation in performance as the group size increased. Petherick et al. (1989) reported that pigs housed in groups of 36 had greater initial variation in BW than groups of 6 or 18 pigs; however, the variation in BW was similar among group sizes when the study concluded.

It has been suggested that one of the main sources of stress in large groups is the complexity of the social hierarchy (Stricklin and Mench, 1987; Moore et al., 1996). If social stress is significantly greater in pens of larger group sizes, one could expect a breakdown in social stability and hence an increase in the variability of animals (Stricklin and Mench, 1987).

One strategy that pigs could use to reduce the complexity would be to form social subgroups that avoid interacting with other subgroups (Stricklin and Mench, 1987; Gonyou and Keeling, 2001). The result of such a strategy would be that the pigs remain in a limited area of the pen and avoid other sections (Schmolke et al., 2003). Additionally, the probability of forming linear hierarchies is relatively low, even for moderate group sizes of seven to eight individuals (Mesterton-Gibbons and Dugatkin, 1995).

The increase in pig weights can lead to increased stocking densities toward the end of the finishing phase, increasing feeder and water competition, causing aggression among pen mates that can lead to an increase in injuries, lameness and other problems.

Resources such as food, water and attractive lying places are often limited in space, creating a competitive environment, which leads to aggression and social stress, even when the amount of resources is sufficient (McGlone, 1985; Mendl et al., 1992). Due to the differences in competitive ability, some individuals will monopolize resources, while others will refrain from competing and use the resources when the other group members are resting (Andersen et al., 2004). Resource monopolization is reported to be reduced in larger groups since a high rate of intrusions reduces the effectiveness of aggression in controlling a resource (Davies and Houston, 1981) and increases the time spent, energy expended and injury (Andersen et al., 2004). As group size increases, more individuals will benefit from not engaging in fights and will use alternative non-aggressive strategies to acquire resources (Andersen et al., 2004). Pagel and Dawkins (1997) suggested that the likeliness of encountering the same individual will decline with increasing group size. However, larger groups imply a more complex social environment, and usually only those individuals with high resource-holding potential are involved in serious fights (Andersen et al., 2004). Thus, it is assumed that there are large individual differences in aggression within groups (Mendl et al., 1992; Hessing et al., 1993; Erhard et al., 1997; Andersen et al., 2000).

5 Living conditions in indoor systems

5.1 Temperature

Confined livestock buildings are usually equipped with mechanical ventilation systems that will (1) provide sufficient air quality during the winter season, in combination with air temperatures that are close to the thermoneutral zone of the animals, despite restricted ventilation rates and (2) removal of sensible heat from animals by high ventilation rates to avoid high house temperatures during the summer (Vitt et al., 2017). The ventilation system must remove water and noxious gases in the barn, but limitations result from economic constraints due to high investment, energy and maintenance costs (Vitt et al., 2017). Newer filtered barns are being developed for increased biosecurity. These barns consume more energy and, if not managed right, can have poor air quality.

Pigs growing near their thermoneutral zone have better feed efficiency and improved welfare (Jackson et al., 2018; Cecchin et al., 2019) compared to when pigs are cold. Pigs in temperatures outside their thermoneutral zone may experience behavioral and physiological changes, consequentially affecting performance (Kiefer et al., 2009). Compared to other livestock species reared for meat production, pigs have a relatively low level of insulation in their coat hair, making them more susceptible to temperatures outside their thermoneutral zone within their housing environment (Jackson et al., 2018). Feral pigs can change their microenvironment by manipulating the material to provide insulation while indoor pigs cannot.

In cold conditions, pigs' thermoregulatory heat production increases, and energy retention from consumed nutrients is depressed, causing worse feed efficiency (Lopez et al., 1991). In warm weather, pigs will begin to reduce feed intake in order to decrease their metabolic heat production (Jackson et al., 2018), leading to a significant impact on pig productivity (Renaudeau et al., 2011).

Pigs have underdeveloped sweat glands, and, thus, it is more likely that pigs will experience heat stress in their housing environment than other farm animals. Pig housing designs should include water misting (Santonja et al., 2017), as this type of cooling system can aid in heat transfer through evaporation and convection (Jackson et al., 2018). Evaporation will transform water droplets into vapor, increasing humidity, reducing the surrounding air temperature, while convection created by an adequate velocity of air passing over the pigs will remove heat and vapor before the pigs get wet (Jackson et al., 2018).

In a study by Vitt et al. (2017), earth heat exchangers (those that use the earth for heat storage) showed the best performance compared to direct evaporative cooling by cooling pads and indirect evaporative cooling by cooling pads in combination with a heat exchanger by totally avoiding heat stress. Earth inlet heat exchangers were effective at damping short-term temperature fluctuations and heating the inlet air temperature during the winter, whereas direct evaporative cooling pads reduced temperature depending on the relative humidity of the outdoor temperature (but moistened bedding materials), and indirect evaporative cooling by cool pads in combination with a heat exchanger reduced the inlet air temperature by evaporation without humidification (Vitt et al., 2017).

5.2 Air quality

Gentry et al. (2002a) studied the lungs of pigs kept indoors on either slats or indoors on bedding. The lungs of pigs kept indoors on bedding, and slats were similar on average. However, pigs in the indoor slatted system had two times the severe lung lesions than pigs on bedding indoors. Air quality is a challenge in both systems; in indoor slatted systems it is ammonia and other gases, while, on bedding, it is dust. Apparently poor gas air quality is worse for pig lungs than the dust from the bedding.

Pig production is one of the major contributors to ammonia production (Groot Koerkamp et al., 1998; Aneja et al., 2000), methane and nitrous oxide (Blanes-Vidal et al., 2008). These gases are produced by direct emission from the digestive system of animals or from the decomposition of animal waste (Blanes-Vidal et al., 2008). Among the different swine buildings, ammonia concentrations are highest in finishing buildings (Groot Koerkamp et al., 1998). Ammonia concentration typically ranges from 0 ppm to 40 ppm (Heber et al., 2005), and the concentration is usually higher at night and in winter when building ventilations are low due to lower outside temperatures (Ni et al., 2018).

Ammonia is toxic and irritates the respiratory tract at concentrations exceeding 15 ppm (Banhazi et al., 2008). High ammonia concentrations can have negative effects (Fig. 4) on human health (Urbain et al., 1994) and the health of pigs (Gerber et al., 1991).

In a study by Jones et al. (1996), pigs were given access to compartments with different ammonia concentrations (0, 10, 20 and 40 ppm). Pigs spent a greater amount of their time in the compartments with lower ammonia concentration than the compartments with higher concentration, 0 ppm (54.3%), 10 ppm (26.9%), 20 ppm (7·1%) and 40 ppm (5.1%), respectively. Pigs visited areas with higher concentrations of ammonia less often and for shorter periods of time. However, this group reported that the aversion to areas with higher

Ammonia concentration (ppm)	Symptoms
10	Some negative effects at long term exposure
15	Smell threshold for human beings
20	Eye irritation for broilers
20–40	Increase of respiratory diseases
25–35	Stockmen feel uncomfortable
50	Disturbance of productive capacity; Water flows from the eyes
50–150	Decrease of young pig growth by 12– 29%
70	Reduced daily gain and poor feed conversion
100–200	Irritation and anorexia
5000	Deadly within a few minutes

Figure 4 Ammonia concentration and the effects/symptoms on humans and animals. Adapted from Ni et al. (2018).

concentrations of ammonia was not immediate and movement to other areas with lower ammonia concentrations may have been due to a sense of malaise that was possibly developed in a polluted atmosphere. The pigs also chose to rest, feed and forage in areas where there was lower ammonia concentration.

5.3 Space allowance

Space allowance is of interest to animal producers, policy makers and animal welfare groups and plays an important role in public concerns associated with intensive farming (Estevez, 2007; SVC, 1996, 1997). Space allowance can affect the welfare and productivity of pigs (Weng et al., 1998; Spoolder et al., 1999). New pig diseases that are endemic in many countries (Chae, 2004; Baekbo et al., 2012) may be exacerbated by social stress and crowding (Sutherland et al., 2007; Alarcon et al., 2011). Space allowance has been extensively studied throughout the years, with a negative effect of high stocking densities being found on different behavioral and physiological indicators associated with impaired welfare (Meunier-Salaun et al., 1987, 2007). Thus, recommendations to producers include reduced stocking density, that is, providing more space (Hassing and Bækbo, 2004).

The formula that relates body weight to surface areas is known as the k-value (NFACC, 2014). When multiplied by a pig's body weight (kg), k-value gives the floor surface area in m^2: $A = k \times BW^{0.667}$, where A = floor surface area in m^2; k-value = floor space allowance coefficient; BW = pig body weight in kg (NFACC, 2014). Allometric space allowance provides a basis for the determination of the spatial needs for pigs (BW; Petherick, 1983; Baxter, 1984) and the homogenization of all the available information according to BW (Gonyou et al., 2006).

In a study by Thomas et al. (2017), different space allowances (0.84, 0.74 or 0.65 m^2 per pig) were tested using finishing pigs. Thomas et al. (2017) found a reduction in ADFI as space allocation decreased, starting at a mean BW of 80.3 kg and that ADFI and ADG decreased linearly, starting at a mean BW of 74 kg, as space allocation decreased. This group concluded that decreasing space allocation resulted in poorer ADG driven by a reduction in ADFI. In another study by Carpenter et al. (2018), over a 71-day period, provided a space allowance of 0.91, or 0.63 m^2 per pig, with either a gate adjustment (on days 28, 45 and 62) or removal of the heaviest pig from the pen (on days 28 and 45) to provide more space and keep pigs in accordance with their predicted minimum space requirement ($(m^2) = 0.0336 \times (BW, kg)^{0.667}$). The results of their study were that pigs provided 0.91 m^2 had increased ADG compared with those allowed 0.63 m^2 with pigs provided space adjustments. Furthermore, pigs with 0.91 m^2 of space grew faster and consumed more feed than pigs restricted in space. As pigs grew to the predicted space requirement and were provided with more space, performance for pigs with adjustments was greater than those provided just 0.63 m² but less than those allowed 0.91 m². Carpenter et al. (2018) concluded that the industry-accepted critical k-value (of 0.0336) may not be adequate for optimal pig performance across multiple BW ranges. Flohr et al. (2018) used data from 30 publications on the effects of floor space allowance on the growth of finishing pigs using alternative prediction equations for ADG, average daily feed intake (ADFI) and gain:feed ratio (G:F). The optimum equations to predict ADG, ADFI and G:F were ADG, g = 337.57 + (16 468 × k) − (237 350 × k2) − (3.1209 × initial BW (kg)) + (2.569 × final BW (kg)) + (71.6918 × k × initial BW (kg)); ADFI, g = 833.41 + (24 785 × k) − (388 998 × k2) − (3.0027 × initial BW (kg)) + (11.246 × final BW (kg)) + (187.61 × k× initial BW (kg)); G:F = predicted ADG/predicted ADFI. The results of the meta-analysis indicated that BW was a predictor of ADG and ADFI even after computing the constant coefficient k, which used the final BW in its calculation. Flohr et al. (2018) suggested that including initial and final BW improves the prediction over using k as a predictor alone and that G:F of finishing pigs is influenced by floor space allowance.

Small space allowances can adversely affect productivity but also pen hygiene (Jensen et al., 2012). Pigs naturally avoid defecating in their lying areas, and generally hygiene will deteriorate as pigs get larger (Hacker et al., 1994; Rossi et al., 2008), as they will not be able to separate their dunning and lying areas with reduced space (Jensen et al., 2012). According to the EFSA the equation to calculate space allowance should be $A = 0.036 \times BW^{0.67}$ for pigs up to 110 kg, but pigs above 110 kg should have a k = 0.047; this equation is based solely on lying behavior and does not incorporate space for fundamental behaviors (such as feeding, drinking, excretion and exploration; Vermeer et al., 2014). A group of international scientists reviewed the scientific literature to determine the k-value for finishing pigs based on the actual published weight gain and other objective measures. Gonyou et al. (2006) concluded the k-value should be between 0.0317 and 0.0348. This range is based on science, whereas the EFSA value is based on the judgment of its authors.

5.4 Lying comfort

Growing-finishing pigs spend a large part of their time budget lying down (Ruckebusch, 1972), and, therefore, an adequate lying comfort seems important for their welfare (Tuyttens, 2005). Data relative to the allometric space required to satisfy pigs' different lying postures already exists (Petherick, 1983; Ekkel et al., 2003), but the quantitative relationships between housing factors and the percentage of time spent lying remain largely unknown (Averós et al., 2010). Averós et al. (2010) conducted a meta-analysis of 22 studies, in which they found a significant interaction between the k-value and the floor type, showing that the relationship between space allowance and lying behavior is affected by the presence or absence of slats. Averós et al. (2010) also suggested that the ability to rest as space availability decreases may be compromised before a reduced performance becomes apparent.

The concept of used and free spaces was first published by McGlone and Newby (1994). First, pigs utilize space differently depending on the time of day. Floor space occupied by standing and lying pigs is least in the middle of the night and most during active periods when more pigs are standing. The authors developed regression equations for free floor space over a day and found the point in which free space was the least. This could be the space requirement – the time when most pigs are lying down. Imagine a pen of pigs sleeping or lying down – the floor space not occupied is the free space. The authors provided experimental evidence for two conclusions. First, considering the USA required space allowance per pig, when 50% of the free space is removed by reducing pen size, then there is no performance set back as there is in crowded pigs. Second, as the group size increased; that is, assume you give heavy finishing pigs 1 m²/pig. With 10 pigs, there is a certain, small amount of free space. But if you give 40 pigs 1 m²/pig, there is a greater amount of free space. With very large group sizes (e.g. 1000 pigs in a pen), the total space can be reduced because there is so much unused, free space. Other authors have not been able to demonstrate this concept using other methods. However, the second point is a simple case of geometry. The first conclusion may work only with very large group sizes (100 or more pigs per pen).

Nannoni et al. (2019) reported that increasing the space allowance to 1.3 m²/head for Italian heavy pigs had positive effects on animal welfare, as it increased the possibility of rest and improved productive parameters and reduced pen floor exploration. When there is insufficient space, group members time-share space; thus the space is dependent not only on space allowance per individual but also on group size (Petherick, 2007). Allowing all the pigs to lay down all at the same time may prevent sleep disruption by other pen mates and allow longer and synchronized sleeping bouts (Nannoni et al., 2019). Synchronized lying behavior has been reported as an indicator of improved welfare (Vermeer et al., 2014) in contrast to Gonyou et al. (2006) who based their conclusion on the point in lowering space allowances in which animal performance was negatively impacted. One cannot know the consequences of changes in lying behavior, but pig feed intake and growth are parameters that are very sensitive to space, especially too little space.

5.5 Lighting

Inappropriate lighting has a negative effect on the welfare of animals, and lighting influences an animal's ocular, physical and neural development and behavior (Taylor et al., 2006). Taylor et al. (2006) placed pigs in four chambers with different illuminance (2.4, 4, 40 and 400 lux) and found that pigs preferred the dimmest lighting and spent the least time in the brightest lighting. Pigs rested and slept in the dimmest lighting and defecated in the brightest lighting. Taylor et al. (2006) suggested that pigs should be provided at least 6 h of dim light (at 2.4 lux). Literature is limited to determine lighting schedules for pigs. Most of the literature reports lighting recommendations for low-stress handling (Grandin, 2003, 2014).

5.6 Stockmanship

The attitude of stockmen/caretakers toward animals is directly associated with their behavior during handling, and rough practices can negatively affect animal welfare (Hemsworth and Coleman, 2010). When animals experience positive interactions with humans, they become less fearful, which, in turn, facilitates handling (Schmied et al., 2010; Probst et al., 2012).

In modern husbandry production, human–animal relationships are an important component of farm animal welfare (Hosey and Melfi, 2014). Poor stockmanship can increase animal anxiety, leading to difficulty in handling and management (Wang et al., 2020). Pigs with a high degree of fear of humans have restricted growth and production performance due to a chronic stress responses (Hemsworth et al., 1981, 1986, 1987; Gonyou et al., 1986). A low-stress response of pigs toward people, and increased willingness to approach and make contact with people is an indicator of good human–animal relationship (Waiblinger et al., 2006). Gentle handling has been reported to improve the daily feed intake (Day et al., 2002), feed conversion rate (Hemsworth and Barnett, 1991) and growth efficiency (Hemsworth et al., 1987) of weaned piglets. Additionally, gentle handling can affect pigs' stress before slaughtering and can lead to improved meat quality (D'Souza et al., 1998; Hemsworth et al., 2002).

Handling practices are usually learned from working 'on the job', and it is common for stockmen to believe that their behaviors are harmless to animals (Hemsworth, 2007). Caretakers' behavior during livestock handling can be improved by carefully selecting personnel and developing training programs designed to reduce animal stress during handling (Boivin et al., 2007). Pig studies have shown that cognitive/behavioral interventions can improve workers' attitudes and consequently their behavior (Coleman et al., 2000; Hemsworth et al., 2002). However, training efficacy depends on the interactions between all trained and non-trained individuals on the same farm and on whether or not a common culture toward animal handling can be developed (Coleman and Hemsworth, 2014).

Coleman et al. (2000) found that training improved stockpeople's attitudes toward animals. In pig production, daily handling often occurs immediately following training sessions, and if done regularly after training, stockpeople can experience the benefits of training by observing the consequences of their behaviors on animals' reactions (Ceballos et al., 2018). Stockpeople's attitude toward the animals could be improved by a feedback loop since the nature and frequency of those behaviors determine the quality of human–animal relationships (Ceballos et al., 2018).

Ceballos et al. (2018) suggests, based on their study, that better managing style within the farm with regular trainings can favor the development of positive stockpeople attitudes toward animals or can reinforce the beneficial effects of training. Additionally, Ceballos et al. (2018) suggested that stockpeople who were not trained but worked on the same team or in direct contact with trained people performed more positive behaviors, which proposes possible social facilitation, imitation or transfer of these types of behaviors. Ceballos et al. (2018) stated that the presence of someone who has been trained has the potential to change the normal context of animal handling on the farm and affects the behavior of the non-trained people, possibly influencing them to behave more properly.

6 Animal health and animal care

Animal health management involves the farm veterinarian, the client or those working for the client/farm caretakers. Early detection of health problems and compromised welfare can increase treatment success, contain problems and enhance pig welfare and sustainability (Matthews et al., 2016). However, early detection of problems is challenging as human observation can be subjective and subclinical disease shows no signs (Matthews et al., 2016).

Inflammatory responses evoke changes in behavior, including increased thermoregulatory activities and sleep, reduced social exploration and appetite and altered food preferences (Millman, 2007). The term 'malaise' is commonly used to describe sickness feelings such as lethargy, depression and pain (Millman, 2007). Expression of sickness behavior would include huddling, shivering, dehydration, fever, anorexia, increased thirst, sleepiness, reduced grooming and exploration, uncoordinated body movements and increased pain and sensitivity (Millman, 2007). Other clinical signs used to identify sick pigs include diarrhea, vomiting, isolating themselves from the social group, seeking warmer environments and increasing lying behavior to minimize heat loss (Millman, 2007). Sickness behavior in pigs may be difficult to identify by the caretaker but also hard for pigs to express, especially toward the end of the finishing phase, where space allowance is reduced due to their group housing environments (group housing, commonly used in commercial settings).

Respiratory disease is also a common problem in late-stage finishing pigs. With increasing body weights, sanitation programs must be taken into consideration as pigs over 110 kg have increased risks of outbreaks of ileitis and mycoplasma pneumonia (Kim et al., 2005). *Mycoplasma hyopneumoniae* is the primary etiologic agent of enzootic pneumonia (a multifactorial disease) of pigs and is a key pathogen in the porcine respiratory disease complex (PRDC) (Thacker and Minion, 2012). *Actinobacillus pleuropneumonia* (APP) is one of the main pathogens associated with pleuritis in pigs (Merialdi et al., 2012). Both of these pathogens are associated with increased body weights, seen toward the end of the finishing phase, and are primarily seen in the USA in larger finishing pigs (Kim et al., 2005). Pigs in the finishing stage are usually affected by these pathogens, which can cause chronic coughing, growth retardation, low mortality but high morbidity (Maes et al., 1996).

6.1 Hospital/sick pens

Properly operated hospital pens (those that offer extra warmth, good footing, easy access to feed and water, have established protocols for treatment, culling and euthanasia) provide an economic return, improve worker morale and improve the welfare of compromised pigs (Blackwell and Wellington Place, 2005). Hospital pens are designed to house compromised pigs where they are supposed to be doctored and nursed back to health, but currently most hospital pens are actually sick pens that are more like holding areas where animals in reality are not commonly treated (Blackwell and Wellington Place, 2005). When a pig dies in a hospital pen, it should always be a surprise. If workers are saying to themselves that they are happy a certain pig finally died, euthanasia decision rules are not adequate and caretakers are neglecting part of their responsibilities (Blackwell and Wellington Place, 2005). As a rule, animals should be culled when they are no longer profitable and euthanized when it is inhumane to let the animal live, but deciding when it is uneconomic and whether to treat or euthanize is difficult for farm managers (Morrow et al., 2006).

The Pork Quality Assurance Plus Program notes that a pig in a hospital pen that does not show signs of improvement after 2 days of intensive care, is severely injured or non-ambulatory with the inability to recover, or any pig that is non-ambulatory with a body condition score (BCS) of 1 should be euthanized.

6.2 Culling

Culling is defined as early pig removal during production (Stein et al., 1990). Culling is important to prevent additional economic losses. The loss of investment at the late stage of production affects the producer's profitability (Jensen et al., 2012). Pigs are typically identified by the caretaker to be sold or are isolated or placed in a separate pen (commonly with other pigs that are compromised known as a sick or hospital pen). Compromised pig segregation allows for their condition to improve by providing them with a nurturing environment that is less competitive, by treating them, or euthanized if the animal does not improve.

6.3 Euthanasia

Subjective guidelines in companion animals look at the ability of the animal to enjoy food, breathe freely and without difficulty, eat and drink without pain and respond to their owner and family (Morrow et al., 2006), whereas objective guidelines evaluate weight loss, weakness, infection, organ failure and injuries (Duncan, 1988). In farm animals, euthanasia considers welfare aspects and economics, but few comprehensive guidelines have been created for swine producers (Morrow et al., 2006). The NPB gives practical recommendations on swine euthanasia, but the On Farm Euthanasia of Swine: Recommendations for the Producer document only describes approved euthanasia methods and not recommendations on etiologies that would require pigs to be euthanized.

Not only is timely euthanasia a matter of good animal welfare, but it is also required by packers and customers. The Common Swine Industry Audit (CSIA) was developed by producers, veterinarians, animal scientists, packers, processors and retail and food-service representatives to establish criteria that any on-farm swine audit must include to be comprehensive and credible (CSIA, 2019). The CSIA includes 27 key aspects of swine care and pre-harvest pork safety through all phases of production. The CSIA tool has five critical questions that, if failed, will cause the farm to fail the entire audit. Question 1 pertains to willful acts of abuse or neglect, and questions 2-5 involve humane euthanasia. Humane euthanasia requires timely euthanasia for animals that have been treated for 2 consecutive days and fail to show signs of improvement. Additional critical questions address the humane handling of animals to be euthanized and the confirmation of insensibility and death. As mentioned earlier, not following these stringent requirements will result in the automatic failure of the audit, leading to a corrective action, which may include being dropped by the packer if repetitive violations occur or the farm fails to meet the required scores. However, despite the criticality of these issues in the industry, neither the CSIA nor Pork Quality Assurance Plus Program recommendations on timely euthanasia have been verified with scientific studies producing empirical evidence.

7 Conclusion and future trends in research

Pig inventory is much greater for pigs from weaning to market than for breeding stock. Therefore, opportunities to improve animal welfare in wean-to-finish pigs are significant if one wants to help many animals. This chapter summarizes areas of animal welfare concern on commercial farms and offers solutions if they are available. The future will bring more automated technologies to improve the health and welfare of pigs.

On many farrow-to-finish farms where 25 pigs per sow per year are typically produced, a 1000-sow farm would have more than 25 000 finishing pigs. Welfare problems in this group impact more individuals than any item in the breeding herd.

In the authors' view, the largest problem in the swine industry is the absence of skilled labor. Many larger farms operate with less than full staff. Having an inadequate number of people directly relates to reduced pig welfare. The solution to this problem is to automate this (and other) stage of production. On many farms, they assign one person to look at eight finishing barns of 1200 pigs each or 9600 pigs in total. This means the worker spends 1 h, with 1200 pigs in a day. That equates to 20 pigs per minute or 3 s per pig, assuming all the time is spent in pig observation – which it is not – workers must check feeders and waters and deal with sick or dead pigs. All of this can be automated using smart barn technologies. Cameras and sensors can monitor temperature, humidity, air flow, air quality, coughs, pig health and pig weights. Algorithms can determine pigs that are out of compliance and alter caretakers to a specific animal to attend to.

One of the most common criticisms of activists is when they find injured or ill pigs. Unattended 'compromised' pigs are common at a low percentage on all farms. Finding these animals quickly and attending to them or conducting timely euthanasia is critical, but when farms are short of labor, it is difficult to get the tasks accomplished. Smart barns may eventually solve these problems.

All the other issues in pig production are less of a problem, but they still warrant attention. These include painful practices, early detection and treatment of pig diseases, space, transport and enrichment. Funding bodies can direct resources toward the most important problems first. One caution is that we should not try to find solutions to problems that are caused by our production system (e.g. tail biting, aggression and others). We should instead change the production system. The finishing system has not changed much in the past 50 years. Current stressors, including concrete slatted floors, no use of bedding, aggression, heat stress, and others merit solutions to prevent the stressor rather than looking for a way to reduce the problem. Prevention is far more effective than treatment (all of which do not work well).

Two other take-home messages are (1) any change in animal welfare for the better must not include losing ground on other society issues (McGlone, 2001) and (2) from an economic point of view, any change to improve welfare cannot increase production costs unless the market pays for the change. Finally, when one country or region implements what they believe as positive welfare changes, it often results in a negative economic outcome. Forcing people to change practices to accommodate views held by other cultures may not be the solution to improving animal welfare.

8 Where to look for further information

The science of pig welfare is a relatively new discipline. While farm animal practices on the farm and at processing plants have been a concern at least since 1906 when Upton Sinclair wrote The Jungle, actual scientific research about animal welfare really started only in the 1960s in Europe and shortly thereafter in North America. Today, swine scientists study pig welfare in every major country that produces and/or consumes pork. The scientific literature about pig (and other species) welfare has grown and continues to grow at a rapid rate. Scientific literature search engines will identify studies published over the last few decades. However, interested readers should also examine publications over the past 200 years or more for information on pig care and husbandry. Even in 1883, Joseph Harris highlighted the importance of good pig care in the second paragraph of his excellent book on the pig (Harris on the Pig, 1883, reprinted 1999 by Lyons Press, NY, NY). The historical literature often provides insights into pig care, especially when production technology is changing.

Government bodies, especially in Europe, have seemingly ever-evolving animal welfare rules. The EU has EU-wide rules about pig welfare, but each member country can have more strict rules about pig care. In the USA and Canada, there are few animal welfare laws, but the industry has published guidelines. Many food companies that purchase pork have animal welfare rules that their suppliers must follow. Guidelines and company rules are used as a basis for auditing pig welfare on farms and at processing plants. While some countries may have government inspectors, this is not the case in the USA. However, detailed on-farm audits are likely to be effective or more effective than government inspectors. In the EU, some countries adhere to EU and country laws via on-farm inspections, while other EU countries do not perform on-farm inspections. Because on-farm audits are required to market pork, when this system is in place, every farm must be audited. Market access requires compliance without any government oversight. Pig trade associations in the USA produce and refine guidelines over time. One can look at the websites of the National Pork Board (www.pork.org) and the National Pork Producers Council (www.nppc.org) for animal welfare information including audit instruments. In addition, the Professional Animal Auditor Association (PAACO; www.animalauditor.org) certifies animal welfare auditors and audit instruments. PAACO provides open access to audit tools.

9 References

Aarnink, A. J. A., Van Den Berg, A. J., Keen, A., Hoeksma, P. and Verstegen, M. W. A. (1996). Effect of slatted floor area on ammonia emission and on the excretory and lying behaviour of growing pigs. *Journal of Agricultural Engineering Research* 64(4), 299-310.

Aarnink, A. J. A., Swierstra, D., Van den Berg, A. J. and Speelman, L. (1997). Effect of type of slatted floor and degree of fouling of solid floor on ammonia emission rates from fattening piggeries. *Journal of Agricultural Engineering Research* 66(2), 93-102.

Alarcon, P., Velasova, M., Mastin, A., Nevel, A., Stärk, K. D. and Wieland, B. (2011). Farm level risk factors associated with severity of post-weaning multi-systemic wasting syndrome. *Preventive Veterinary Medicine* 101(3-4), 182-191.

Andersen, I. L. and Bøe, K. E. (1999). Straw bedding or concrete floor for loose-housed pregnant sows: consequences for aggression, production and physical health. *Acta Agriculturae Scandinavica, Section A - Animal Science* 49(3), 190-195.

Andersen, I. L., Andenæs, H., Bøe, K. E., Jensen, P. and Bakken, M. (2000). The effects of weight asymmetry and resource distribution on aggression in groups of unacquainted pigs. *Applied Animal Behaviour Science* 68(2), 107-120.

Andersen, I. L., Nævdal, E., Bakken, M. and Bøe, K. E. (2004). Aggression and group size in domesticated pigs, Sus scrofa: 'when the winner takes it all and the loser is standing small'. *Animal Behaviour* 68(4), 965-975.

Aneja, V. P., Chauhan, J. P. and Walker, J. T. (2000). Characterization of atmospheric ammonia emissions from swine waste storage and treatment lagoons. *Journal of Geophysical Research: Atmospheres* 105(D9), 11535-11545.

Anil, L., Anil, S. S., Deen, J., Baidoo, S. K. and Walker, R. D. (2006). Effect of group size and structure on the welfare and performance of pregnant sows in pens with electronic sow feeders. *Canadian Journal of Veterinary Research = Revue Canadienne de Recherche Vétérinaire* 70(2), 128-136.

Appleby, M. C. (2004). Alternatives to conventional livestock production methods. In: Benson, G. J. and Rollin, B. E. (Eds) *The Well-Being of Farm Animals*, pp. 339-350. John Wiley & Sons.

Araújo, J. P., Amorim, I., Silva, J. S., Pires, P. and Cerqueira, J. (2016). Outdoor housing systems for Bísaro pig breed with a hoop barn: some effects on welfare. In: *Food Futures: Ethics, Science and Culture*, pp. 185-242. Wageningen Academic Publishers, Wageningen, The Netherlands.

Arey, D. S. and Edwards, S. A. (1998). Factors influencing aggression between sows after mixing and the consequences for welfare and production. *Livestock Production Science* 56(1), 61-70.

Averós, X., Brossard, L., Dourmad, J. Y., De Greef, K. H., Edge, H. L., Edwards, S. A. and Meunier-Salaün, M. C. (2010). Quantitative assessment of the effects of space allowance, group size and floor characteristics on the lying behaviour of growing-finishing pigs. *Animal: an International Journal of Animal Bioscience* 4(5), 777-783.

Baekbo, P., Kristensen, C. S. and Larsen, L. E. (2012). Porcine circovirus diseases: a review of PMWS. *Transboundary and Emerging Diseases* 59(Suppl. 1), 60-67.

Bai, Z., Ma, W., Ma, L., Velthof, G. L., Wei, Z., Havlík, P., Oenema, O., Lee, M. R. F. and Zhang, F. (2018). China's livestock transition: driving forces, impacts, and consequences. *Science Advances* 4(7), eaar8534.

Banhazi, T. M., Seedorf, J., Rutley, D. L. and Pitchford, W. S. (2008). Identification of risk factors for sub-optimal housing conditions in Australian piggeries: Part 2. Airborne pollutants. *Journal of Agricultural Safety and Health* 14(1), 21–39.

Baxter, S. (1984). *Intensive Pig Production: Environmental Management and Design*. Granada Technical Books, London.

Beattie, V. E., Sneddon, I. A., Walker, N. and Weatherup, R. N. (2001). Environmental enrichment of intensive pig housing using spent mushroom compost. *Animal Science* 72(1), 35–42.

Bee, G., Guex, G. and Herzog, W. (2004). Free-range rearing of pigs during the winter: adaptations in muscle fiber characteristics and effects on adipose tissue composition and meat quality traits. *Journal of Animal Science* 82(4), 1206–1218.

Blackwell, T. and Wellington Place, R. R. (2005). Effective treatment and handling of poor doing pigs in the finishing barn. In: Conference Proceedings, London Swine Conference, London, Ontario, p. 167.

Blanes-Vidal, V., Hansen, M. N., Pedersen, S. and Rom, H. B. (2008). Emissions of ammonia, methane and nitrous oxide from pig houses and slurry: effects of rooting material, animal activity and ventilation flow. *Agriculture, Ecosystems and Environment* 124(3–4), 237–244.

Bolhuis, J. E., Schouten, W. G. P., Schrama, J. W. and Wiegant, V. M. (2005). Behavioural development of pigs with different coping characteristics in barren and substrate-enriched housing conditions. *Applied Animal Behaviour Science* 93(3-4), 213–228.

Boivin, X., Marcantognini, L., Boulesteix, P., Godet, J., Brulé, A. and Veissier, I. (2007). Attitudes of farmers towards Limousin cattle and their handling. *Animal Welfare Potters Bar Then Whethampstead* 16(2), 147.

Broom, D. M. (2010). Animal welfare: an aspect of care, sustainability, and food quality required by the public. *Journal of Veterinary Medical Education* 37(1), 83–88.

Broom, D. M., Gunn, M., Edwards, S., Wechsler, B., Algers, B., Spoolder, H., Madec, F., Von Borell, E. and Olsson, O. (2005). The Welfare of Weaners and Rearing Pigs: Effects of Different Space Allowances and Floor Types. EFSA-Q-2004-077, p. 129.

Brumm, M. (2012). Impact of heavy market weights on facility and equipment needs. Proceedings of the Allen D. Leman Swine Conference. St Paul, Minnesota, pp. 165–168.

Brumm, M. C., Harmon, J. D., Honeyman, M. S., Kliebenstein, J. B., Lonergan, S. M., Morrison, R. and Richard, T. (2004). *Hoop Barns for Grow-Finish Swine*. MidWest Plan Service, Ames, IA.

Calderón Díaz, J. A., Diana, A., Boyle, L. A., Leonard, F. C., McElroy, M., McGettrick, S., Moriarty, J. and García Manzanilla, E. (2017). Delaying pigs from the normal production flow is associated with health problems and poorer performance. *Porcine Health Management* 3(1), 13.

Caley, P. (1997). Movements, activity patterns and habitat use of feral pigs (Sus scrofa) in a tropical habitat. *Wildlife Research* 24(1), 77–87.

Callaway, T. R., Morrow, J. L., Johnson, A. K., Dailey, J. W., Wallace, F. M., Wagstrom, E. A., McGlone, J. J., Lewis, A. R., Dowd, S. E., Poole, T. L., Edrington, T. S., Anderson, R. C., Genovese, K. J., Byrd, J. A., Harvey, R. B. and Nisbet, D. J. (2005). Environmental prevalence and persistence of Salmonella spp. in outdoor swine wallows. *Foodborne Pathogens and Disease* 2(3), 263–273.

Cameron, R. (2012). Integumentary system, skin, hoof and claw. In Zimmerman, J. J. (Ed.), *Diseases of Swine*, pp. 251–263. Wiley-Blackwell, West Sussex, UK.

Carpenter, C. B., Holder, C. J., Wu, F., Woodworth, J. C., DeRouchey, J. M., Tokach, M. D., Goodband, R. D. and Dritz, S. S. (2018). Effects of increasing space allowance by removing a pig or gate adjustment on finishing pig growth performance. *Journal of Animal Science* 96(7), 2659-2664.

Casal-Plana, N., Manteca, X., Dalmau, A. and Fàbrega, E. (2017). Influence of enrichment material and herbal compounds in the behaviour and performance of growing pigs. *Applied Animal Behaviour Science* 195, 38-43.

Ceballos, M. C., Sant'Anna, A. C., Boivin, X., Costa, FdO., Carvalhal, M. V. and Paranhos da Costa, M. J. R. (2018). Impact of good practices of handling training on beef cattle welfare and stockpeople attitudes and behaviors. *Livestock Science* 216, 24-31.

Cecchin, D., Ferraz, P. F. P., Campos, A. T., Sousa, F. A., Amaral, P. I. S., Castro, J. O., Conti, L. and da Cruz, V. M. F. (2019). Thermal comfort of pigs housed in different installations. *Agronomy Research* 17(2), 378-384.

Chae, C. (2004). Postweaning multisystemic wasting syndrome: a review of aetiology, diagnosis and pathology. *The Veterinary Journal* 168(1), 41-49.

Clark, B., Panzone, L. A., Stewart, G. B., Kyriazakis, I., Niemi, J. K., Latvala, T., Tranter, R., Jones, P. and Frewer, L. J. (2019). Consumer attitudes towards production diseases in intensive production systems. *PLoS ONE* 14(1), e0210432.

Coleman, G. J., Hemsworth, P. H., Hay, M. and Cox, M. (2000). Modifying stockperson attitudes and behaviour towards pigs at a large commercial farm. *Applied Animal Behaviour Science* 66(1-2), 11-20.

Coleman, G. J. and Hemsworth, P. H. (2014). Training to improve stockperson beliefs and behaviour towards livestock enhances welfare and productivity. *Revue Scientifique et Technique* 33(1), 131-137.

Common Swine Industry Audit (2019). Available at: https://www.pork.org/production/tools/common-swine-industry-audit/common-swine-industry-audit-materials/.

Cooper, J. J., Cox, L. N. and Whitworth, C. (2001). Early environmental experience and transferable skills in the weaned piglet. *Animal Welfare-Potters Bar* 10, S238-S238.

Dailey, J. W. and McGlone, J. J. (1997). Oral/nasal/facial and other behaviors of sows kept individually outdoors on pasture, soil or indoors in gestation crates. *Applied Animal Behaviour Science* 52(1-2), 25-43.

Davies, N. B. and Houston, A. I. (1981). Owners & satellites: the economics of territory defence in the pied wagtail, Motacilla alba. *Journal of Animal Ecology* 50(1), 157-180.

Day, J. E. L., Spoolder, H. A. M., Burfoot, A., Chamberlain, H. L. and Edwards, S. A. (2002). The separate and interactive effects of handling and environmental enrichment on the behaviour and welfare of growing pigs. *Applied Animal Behaviour Science* 75(3), 177-192.

Devillers, N., Janvier, E., Delijani, F., Méthot, S., Dick, K. J., Zhang, Q. and Connor, L. (2019). Effect of slat and gap width of slatted concrete flooring on sow gait using kinematics analysis. *Animals: an Open Access Journal from MDPI* 9(5), 206.

Diana, A., Boyle, L. A., García Manzanilla, E., Leonard, F. C. and Calderón Díaz, J. A. (2019). Ear, tail and skin lesions vary according to different production flows in a farrow-to-finish pig farm. *Porcine Health Management* 5(1), 19.

Dixon, L. K., Sun, H. and Roberts, H. (2019). African swine fever. *Antiviral Research* 165, 34-41.

Docking, C. M., Van de Weerd, H. A., Day, J. E. L. and Edwards, S. A. (2008). The influence of age on the use of potential enrichment objects and synchronisation of behaviour of pigs. *Applied Animal Behaviour Science* 110(3-4), 244-257.

D'Souza, D. N., Leury, B. J., Dunshea, F. R. and Warner, R. D. (1998). Effect of on-farm and pre-slaughter handling of pigs on meat quality. *Australian Journal of Agricultural Research* 49(6), 1021-1025.

Duncan, J. C. (1988). Careers in veterinary medicine. New York, New York: Rosen Publishing Group.

EFSA Panel on Animal Health and Welfare (2007). Scientific opinion of the Panel on Animal Health and Welfare on a request from Commission on the Risks Associated with Tail Biting in Pigs and Possible Means. *The EFSA Journal* 611, 1-13.

EFSA Panel on Animal Health and Welfare (AHAW) (2012). Statement on the use of animal-based measures to assess the welfare of animals. *EFSA Journal* 10(6), 2767.

EFSA Panel on Animal Health and Welfare (AHAW) (2014). Scientific opinion concerning a multifactorial approach on the use of animal and non-animal-based measures to assess the welfare of pigs. *EFSA Journal* 12(5), 3702.

Einarsson, S., Madej, A. and Tsuma, V. (1996). The influence of stress on early pregnancy in the pig. *Animal Reproduction Science* 42(1-4), 165-172.

Ekkel, E. D., Spoolder, H. A. M., Hulsegge, I. and Hopster, H. (2003). Lying characteristics as determinants for space requirements in pigs. *Applied Animal Behaviour Science* 80(1), 19-30.

English, P. R. (1988). *The Growing and Finishing Pig: Improving Efficiency (No. SF 396.9. G76)*.

Erhard, H. W., Mendl, M. and Ashley, D. D. (1997). Individual aggressiveness of pigs can be measured and used to reduce aggression after mixing. *Applied Animal Behaviour Science* 54(2-3), 137-151.

Estevez, I. (2007). Density allowances for broilers: where to set the limits? *Poultry Science* 86(6), 1265-1272.

Falke, A., Friedli, K., Gygax, L., Wechsler, B., Sidler, X. and Weber, R. (2018). Effect of rubber mats and perforation in the lying area on claw and limb lesions of fattening pigs. *Animal: an International Journal of Animal Bioscience* 12(10), 2130-2137.

Flohr, J. R., Dritz, S. S., Tokach, M. D., Woodworth, J. C., DeRouchey, J. M. and Goodband, R. D. (2018). Development of equations to predict the influence of floor space on average daily gain, average daily feed intake and gain: feed ratio of finishing pigs. *Animal: an International Journal of Animal Bioscience* 12(5), 1022-1029.

Fraser, D. (1978). Observations on the behavioural development of suckling and early-weaned piglets during the first six weeks after birth. *Animal Behaviour* 26, 22-30.

Fraser, D. G. (2005). *Animal Welfare and the Intensification of Animal Production: An Alternative Interpretation*, Vol. 2. Food & Agriculture Org.

Frei, D., Würbel, H., Wechsler, B., Gygax, L., Burla, J. B. and Weber, R. (2018). Can body nosing in artificially reared piglets be reduced by sucking and massaging dummies? *Applied Animal Behaviour Science* 202, 20-27.

Fritschen, R. D. and Hogg, A. (1996). *Preventing Tail Biting in Swine (Anti-Comfort Syndrome)*. Cooperative Extension, Institute of Agriculture and Natural Resources, University of Nebraska--Lincoln.

Gardner, J. M., De Lange, C. F. M. and Widowski, T. M. (2001a). Belly-nosing in early-weaned piglets is not influenced by diet quality or the presence of milk in the diet. *Journal of Animal Science* 79(1), 73-80.

Gardner, J. M., Duncan, I. J. H. and Widowski, T. M. (2001b). Effects of social "stressors" on belly-nosing behaviour in early-weaned piglets: is belly-nosing an indicator of stress? *Applied Animal Behaviour Science* 74(2), 135-152.

Gentry, J. G., McGlone, J. J., Blanton, J. R. and Miller, M. F. (2002a). Alternative housing systems for pigs: influences on growth, composition and pork quality. *Journal of Animal Science* 80(7), 1781-1790.

Gentry, J. G., McGlone, J. J., Blanton, J. R., Jr. and Miller, M. F. (2002b). Impact of spontaneous exercise on performance, meat quality, and muscle fiber characteristics of growing/finishing pigs. *Journal of Animal Science* 80(11), 2833-2839.

Gentry, J. G., McGlone, J. J., Miller, M. F. and Blanton, J. R., Jr. (2004). Environmental effects on pig performance, meat quality, and muscle characteristics. *Journal of Animal Science* 82(1), 209-217.

Gerber, D. B., Veenhuizen, M. A. and Shurson, G. C. (1991). Ammonia, carbon monoxide, carbon dioxide, hydroden sulfide, and methane in swine confinement facilities. *The Compendium on Continuing Education for the Practicing Veterinarian (USA)* 13, 1483-1488.

Gondret, F., Combes, S., Lefaucher, L. and Lebret, B. (2005). Effects of exercise during growth and alternative rearing systems on muscle fibers and collagen properties. *Reproduction, Nutrition, Development* 45(1), 69-86.

Gonyou, H. W., Hemsworth, P. H. and Barnett, J. L. (1986). Effects of frequent interactions with humans on growing pigs. *Applied Animal Behaviour Science* 16(3), 269-278.

Gonyou, H. W. and Keeling, L. J. (Eds) (2001). *Social Behavior in Farm Animals*. CABI Publishing, Wallingford, UK.

Gonyou, H. W., Brumm, M. C., Bush, E., Davies, P., Deen, J., Edwards, S. A., Fangman, T., McGlone, J. J., Meunier-Salaun, M., Morrison, R. B., Spoolder, H., Sundberg, P. L. and Johnson, A. K. (2006). Application of broken line analysis to assess floor space requirements of nursery and grow/finish pigs expressed on a allometric basis. *Journal of Animal Science* 84, 229-235.

Grandin, T. (2003). Transferring results of behavioral research to industry to improve animal welfare on the farm, ranch and the slaughter plant. *Applied Animal Behaviour Science* 81(3), 215-228.

Grandin, T. (2014). Animal welfare and society concerns finding the missing link. *Meat Science* 98(3), 461-469.

Grandin, T. and Whiting, M. (Eds) (2018). *Are We Pushing Animals to Their Biological Limits?: Welfare and Ethical Implications*. CABI Publishing, Wallingford, UK.

Groot Koerkamp, P. W. G., Metz, J. H. M., Uenk, G. H., Phillips, V. R., Holden, M. R., Sneath, R. W., Short, J. L., White, R. P. P., Hartung, J., Seedorf, J., Schröder, M., Linkert, K. H., Pedersen, S., Takai, H., Johnsen, J. O. and Wathes, C. M. (1998). Concentrations and emissions of ammonia in livestock buildings in northern Europe. *Journal of Agricultural Engineering Research* 70(1), 79-95.

Guy, J. H., Rowlinson, P., Chadwick, J. P. and Ellis, M. (2002). Health conditions of two genotypes of growing-finishing pig in three different housing systems: implications for welfare. *Livestock Production Science* 75(3), 233-243.

Hacker, R. R., Ogilvie, J. R., Morrison, W. D. and Kains, F. (1994). Factors affecting excretory behavior of pigs. *Journal of Animal Science* 72(6), 1455-1460.

Hassing, A.-G. and Bækbo, P. (2004). PMWS manual. Videncenter for Svineproduction. Centre for Pig Production, Copenhagen, Denmark. Available at: http://vsp.lf.dk/Pub likationer/Manualer/PMWS%20manual.aspx?450 full=1.

Heber, A. J., Tao, P. C., Ni, J. Q., Lim, T. T. and Schmidt, A. M. (2005). Air emissions from two swine finishing building with flushing: ammonia characteristics. In: *Livestock*

Environment VII, 18-20 May 2005, Beijing, China, p. 436. American Society of Agricultural and Biological Engineers.

Hemsworth, P. H. (2007). Ethical stockmanship. *Australian Veterinary Journal* 85(5), 194-200.

Hemsworth, P. H. and Barnett, J. L. (1991). The effects of aversively handling pigs, either individually or in groups, on their behaviour, growth and corticosteroids. *Applied Animal Behaviour Science* 30(1-2), 61-72.

Hemsworth, P. H. and Coleman, G. J. (2010). *Human-Livestock Interactions: The Stockperson and the Productivity of Intensively Farmed Animals*. CABI Publishing, Wallingford, UK.

Hemsworth, P. H., Barnett, J. L. and Hansen, C. (1981). The influence of handling by humans on the behavior, growth, and corticosteroids in the juvenile female pig. *Hormones and Behavior* 15(4), 396-403.

Hemsworth, P. H., Barnett, J. L. and Hansen, C. (1986). The influence of handling by humans on the behaviour, reproduction and corticosteroids of male and female pigs. *Applied Animal Behaviour Science* 15(4), 303-314.

Hemsworth, P. H., Barnett, J. L. and Hansen, C. (1987). The influence of inconsistent handling by humans on the behaviour, growth and corticosteroids of young pigs. *Applied Animal Behaviour Science* 17(3-4), 245-252.

Hemsworth, P. H., Barnett, J. L., Hofmeyr, C., Coleman, G. J., Dowling, S. and Boyce, J. (2002). The effects of fear of humans and pre-slaughter handling on the meat quality of pigs. *Australian Journal of Agricultural Research* 53(4), 493-501.

Hemsworth, P. H., Mellor, D. J., Cronin, G. M. and Tilbrook, A. J. (2015). Scientific assessment of animal welfare. *New Zealand Veterinary Journal* 63(1), 24-30.

Hessing, M. J. C., Hagelsø, A. M., van Beek, J. A. M., Wiepkema, R. P., Schouten, W. G. P. and Krukow, R. (1993). Individual behavioural characteristics in pigs. *Applied Animal Behaviour Science* 37(4), 285-295.

Hill, J. (2000). Deep bed swine finishing. *5 o Seminário Internacional de Suinocultura 27 e 28 de setembro de 2000*. Expo Center Norte, SP. Available at: http://www.cnpsa .embrapa.br/sgc/sgc_publicacoes/anais0009_hill.pdf.

Honeyman, M. S. (2005). Extensive bedded indoor and outdoor pig production systems in USA: current trends and effects on animal care and product quality. *Livestock Production Science* 94(1-2), 15-24.

Hosey, G. and Melfi, V. (2014). Human-animal interactions, relationships and bonds: a review and analysis of the literature. *International Journal of Comparative Psychology* 27(1), 117-142.

Jackson, P., Guy, J. H., Sturm, B., Bull, S. and Edwards, S. A. (2018). An innovative concept building design incorporating passive technology to improve resource efficiency and welfare of finishing pigs. *Biosystems Engineering* 174, 190-203.

Jensen, P., Von Borell, E., Broom, D. M., Csermely, D., Dijkhuizen, A. A., Hylkema, S., Edwards, S. A., Madec, F. and Stamataris, C. (1997). The Welfare of Intensively Kept Pigs. Doc XXIV/B3/ScVC/0005/1997, p. 190. Commission of the European Communities; Scientific Veterinary Committee, Brussels, Belgium.

Jensen, T., Nielsen, C. K., Vinther, J. and D'Eath, R. B. (2012). The effect of space allowance for finishing pigs on productivity and pen hygiene. *Livestock Science* 149(1-2), 33-40.

Jones, J. B., Burgess, L. R., Webster, A. J. F. and Wathes, C. M. (1996). Behavioural responses of pigs to atmospheric ammonia in a chronic choice test. *Animal Science* 63(3), 437-445.

Jørgensen, B. (2003). Influence of floor type and stocking density on leg weakness, osteochondrosis and claw disorders in slaughter pigs. *Animal Science* 77(3), 439-449.

Kiefer, C., Meignen, B. C. G., Sanches, J. F. and Carrijo, E. A. S. (2009). Response of growing swine maintained in different thermal environments. *Archivos de Zootecnia* 58(221), 55-64.

Kim, Y. S., Kim, S. W., Weaver, M. A. and Lee, C. Y. (2005). Increasing the pig market weight: world trends, expected consequences and practical considerations. *Asian-Australasian Journal of Animal Sciences* 18(4), 590-600.

Kittawornrat, A. and Zimmerman, J. J. (2011). Toward a better understanding of pig behavior and pig welfare. *Animal Health Research Reviews* 12(1), 25-32.

Kliebenstein, J., Larson, B., Honeyman, M. and Penner, A. (2003). *A Comparison of Production Costs, Returns and Profitability of Swine Finishing Systems*. Iowa State University, Ames, IA.

Knox, R., Salak-Johnson, J., Hopgood, M., Greiner, L. and Connor, J. (2014). Effect of day of mixing gestating sows on measures of reproductive performance and animal welfare. *Journal of Animal Science* 92(4), 1698-1707.

Langbein, J. and Puppe, B. (2004). Analysing dominance relationships by sociometric methods–a plea for a more standardised and precise approach in farm animals. *Applied Animal Behaviour Science* 87(3-4), 293-315.

Larson, B., Kliebenstein, J. B., Honeyman, M. S. and Penner, A. D. (2002). An economic analysis of pork production in hoop and confinement facilities: a winter comparison. *Animal Industry Report* 1(1).

Larson, B., Kliebenstein, J. B., Honeyman, M. S. and Penner, A. (2003). Economics of finishing pigs in hoop structures and confinement: a summer group under different space restrictions. *Animal Industry Report* 1(1).

Lassaletta, L., Billen, G., Romero, E., Garnier, J. and Aguilera, E. (2014). How changes in diet and trade patterns have shaped the N cycle at the national scale: Spain (1961-2009). *Regional Environmental Change* 14(2), 785-797.

Lassaletta, L., Estellés, F., Beusen, A. H. W., Bouwman, L., Calvet, S., Van Grinsven, H. J. M., Doelman, J. C., Stehfest, E., Uwizeye, A. and Westhoek, H. (2019). Future global pig production systems according to the Shared Socioeconomic Pathways. *Science of the Total Environment* 665, 739-751.

Lay Jr., D. C., Haussmann, M. F., Daniels, M. J., Harmon, J. D. and Richard, T. L. (2000). Swine housing impacts on environment and behavior: a comparison between hoop structures and total environmental control. In: *Swine Housing, Proc. First Int. Conf.* (October 9-11, 2000, Des Moines, Iowa), pp. 49-55. ASAE. Pub., St. Joseph, MI.

Li, Y. and Gonyou, H. W. (2002). Analysis of belly nosing and associated behaviour among pigs weaned at 12-14 days of age. *Applied Animal Behaviour Science* 77(4), 285-294.

Li, Y. Z., Wang, L. H. and Johnston, L. J. (2012). Sorting by parity to reduce aggression toward first-parity sows in group-gestation housing systems. *Journal of Animal Science* 90(12), 4514-4522.

Lopez, J., Jesse, G. W., Becker, B. A. and Ellersieck, M. R. (1991). Effects of temperature on the performance of finishing swine: I. Effects of a hot, diurnal temperature on average daily gain, feed intake, and feed efficiency. *Journal of Animal Science* 69(5), 1843-1849.

Maes, D., Verdonck, M., Deluyker, H. and de Kruif, A. (1996). Enzootic pneumonia in pigs. *Veterinary Quarterly* 18(3), 104-109.

Manteca, X., Velarde, A. and Jones, B. (2009). Animal welfare components. In: Smulders, F. and Algers, B. (Eds) *Welfare of Production Animals: Assessment and Management of Risks*, pp. 61-77. Wageningen Academic Publishers, The Netherlands.

Marchant-Forde, J. N. and Marchant-Forde, R. M. (2005). Minimizing inter-pig aggression during mixing. *Pig News and Information* 26(3), 63N.

Martin, L. and Kruja, Z. (2000). The western Canada advantage. *Advances in Pork Production* 11, 17-36.

Martínez-Miró, S., Tecles, F., Ramón, M., Escribano, D., Hernández, F., Madrid, J., Orengo, J., Martínez-Subiela, S., Manteca, X. and Cerón, J. J. (2016). Causes, consequences and biomarkers of stress in swine: an update. *BMC Veterinary Research* 12(1), 171.

Matthews, S. G., Miller, A. L., Clapp, J., Plötz, T. and Kyriazakis, I. (2016). Early detection of health and welfare compromises through automated detection of behavioural changes in pigs. *The Veterinary Journal* 217, 43-51.

McGlone, J. J. (1985). A quantitative ethogram of aggressive and submissive behaviors in recently regrouped pigs. *Journal of Animal Science* 61(3), 559-565.

McGlone, J. J. (1986). Influence of resources on pig aggression and dominance. *Behavioural Processes* 12(2), 135-144.

McGlone, J. J. (2001). Farm animal welfare in the context of other society issues: toward sustainable systems. *Livestock Production Science* 72(1-2), 75-81.

McGlone, J. J. and Curtis, S. E. (1985). Behavior and performance of weanling pigs in pens equipped with hide areas. *Journal of Animal Science* 60(1), 20-24.

McGlone, J. J. and Newby, B. E. (1994). Space requirements for finishing pigs in confinement: behavior and performance while group size and space vary. *Applied Animal Behaviour Science* 39(3-4), 331-338.

McGlone, J. J., Kelley, K. W. and Gaskins, C. T. (1980). Lithium and porcine aggression. *Journal of Animal Science* 51(2), 447-455.

Mendl, M., Zanella, A. J. and Broom, D. M. (1992). Physiological and reproductive correlates of behavioural strategies in female domestic pigs. *Animal Behaviour* 44(6), 1107-1121.

Merialdi, G., Dottori, M., Bonilauri, P., Luppi, A., Gozio, S., Pozzi, P., Spaggiari, B. and Martelli, P. (2012). Survey of pleuritis and pulmonary lesions in pigs at abattoir with a focus on the extent of the condition and herd risk factors. *The Veterinary Journal* 193(1), 234-239.

Mesterton-Gibbons, M. and Dugatkin, L. A. (1995). Toward a theory of dominance hierarchies: effects of assessment, group size, and variation in fighting ability. *Behavioral Ecology* 6(4), 416-423.

Metz, J. H. M. and Gonyou, H. W. (1990). Effect of age and housing conditions on the behavioural and haemolytic reaction of piglets to weaning. *Applied Animal Behaviour Science* 27(4), 299-309.

Meunier-Salaun, M. C., Vantrimponte, M. N., Raab, A. and Dantzer, R. (1987). Effect of floor area restriction upon performance, behavior and physiology of growing-finishing pigs. *Journal of Animal Science* 64(5), 1371-1377.

Meunier-Salaun, M. C., Bizeray, D., Colson, V., Courboulay, V., Lensink, J., Prunier, A., Remience, V. and Vandenheede, M. (2007). The welfare of farmed pigs. *Productions Animales* 20, 73-80.

Miele, M. and Evans, A. (2010). When foods become animals: ruminations on ethics and responsibility in care-full practices of consumption. *Ethics, Place and Environment* 13(2), 171-190.

Millman, S. T. (2007). Sickness behaviour and its relevance to animal welfare assessment at the group level. *Animal Welfare* 16, 123-125.

Moore, C. M., Zhou, J. Z., Stricklin, W. R. and Gonyou, H. W. (1996). Influence of group size and floor area space on social organization of growing-finishing pigs. In: Proceedings of the 30th International Congress of the International Society for Applied Ethology, 14-17 August, 1996, Guelph, Ontario, Canada. Colonel KL Campbell Centre for the Study of Animal Welfare, Guelph, Ontario.

Mormède, P., Andanson, S., Aupérin, B., Beerda, B., Guémené, D., Malmkvist, J., Manteca, X., Manteuffel, G., Prunet, P., van Reenen, C. G., Richard, S. and Veissier, I. (2007). Exploration of the hypothalamic–pituitary–adrenal function as a tool to evaluate animal welfare. *Physiology and Behavior* 92(3), 317-339.

Morrow, W. M., Meyer, R. E., Roberts, J. and Lascelles, D. (2006). Financial and welfare implications of immediately euthanizing compromised nursery pigs. *Journal of Swine Health and Production* 14(1), 25.

Nannoni, E., Martelli, G., Rubini, G. and Sardi, L. (2019). Effects of increased space allowance on animal welfare, meat and ham quality of heavy pigs slaughtered at 160Kg. *PLoS ONE* 14(2), e0212417.

Nannoni, E., Sardi, L., Vitali, M., Trevisi, E., Ferrari, A., Barone, F., Bacci, M. L., Barbieri, S. and Martelli, G. (2016). Effects of different enrichment devices on some welfare indicators of post-weaned undocked piglets. *Applied Animal Behaviour Science* 184, 25-34.

Nasirahmadi, A., Edwards, S. A. and Sturm, B. (2017). Implementation of machine vision for detecting behaviour of cattle and pigs. *Livestock Science* 202, 25-38.

NASS (2017). Hogs: commercial slaughter average liveweight by month and year, US. Available at: https://www.nass.usda.gov/Charts_and_Maps/Livestock_Slaughter/hglvwgx6.php.

National Farm Animal Care Council (NFACC) (2014). Available at: https://www.nfacc.ca/pdfs/codes/pig_code_of_practice.pdf.

National Pork Board (2019). Available at: https://www.pork.org/facts/pig-farming/life-cycle-of-a-market-pig/.

Ni, J. Q., Heber, A. J. and Lim, T. T. (2018). Ammonia and hydrogen sulfide in swine production. In: *Air Quality and Livestock Farming*, pp. 69-88. CRC Press, London, UK.

OIE (2010). Available at: https://www.oie.int/en/animal-welfare/animal-welfare-at-a-glance/.

Owsley, F., Rodning, S. and Floyd, J. (2013). *Scheduling All-In/All-Out Swine Production, vol ANR-0847. Alabama Cooperative Extension System*.

Pagel, M. and Dawkins, M. S. (1997). Peck orders and group size in laying hens:futures contracts' for non-aggression. *Behavioural Processes* 40(1), 13-25.

Penny, R. H., Osborne, A. D., Wright, A. I. and Stephens, T. K. (1965). Foot-rot in pigs, observations on the clinical disease. *The Veterinary Record* 77(38), 1101-1108.

Petherick, J. C. (1983). A biological basis for the design of space in livestock housing. *Farm Animal Housing and Welfare*, 103-120.

Petherick, J. C. (2007). Spatial requirements of animals: allometry and beyond. *Journal of Veterinary Behavior* 2(6), 197-204.

Petherick, J. C., Beattie, A. W. and Bodero, D. A. V. (1989). The effect of group size on the performance of growing pigs. *Animal Science* 49(3), 497-502.

Probst, J. K., Neff, A. S., Leiber, F., Kreuzer, M. and Hillmann, E. (2012). Gentle touching in early life reduces avoidance distance and slaughter stress in beef cattle. *Applied Animal Behaviour Science* 139(1-2), 42-49.

Rahse, E. and Hoy, S. (2007). Investigations on frequency and severity of different claw lesions in fattening pigs with regard to housing conditions. *Praktische Tierarzt* 88(1), 40-47.

Rault, J. L., Morrison, R. S., Hansen, C. F., Hansen, L. U. and Hemsworth, P. H. (2014). Effects of group housing after weaning on sow welfare and sexual behavior. *Journal of Animal Science* 92(12), 5683-5692.

Reicks, D. (2009). Application of air filtration systems in swine operations. *Advances in Pork Production* 20, 163-171.

Renaudeau, D., Gourdine, J. L. and St-Pierre, N. R. (2011). A meta-analysis of the effects of high ambient temperature on growth performance of growing-finishing pigs. *Journal of Animal Science* 89(7), 2220-2230.

Rossi, R., Costa, A., Guarino, M., Laicini, F., Pastorelli, G. and Corino, C. (2008). Effect of group size-floor space allowance and floor type on growth performance and carcass characteristics of heavy pigs. *Journal of Swine Health and Production* 16(6), 304-311.

Ruckebusch, Y. (1972). The relevance of drowsiness in the circadian cycle of farm animals. *Animal Behaviour* 20(4), 637-643.

Rzeszniczek, M., Gygax, L., Wechsler, B. and Weber, R. (2015). Comparison of the behaviour of piglets raised in an artificial rearing system or reared by the sow. *Applied Animal Behaviour Science* 165, 57-65.

Sánchez-Vizcaíno, J. M., Laddomada, A. and Arias, M. L. (2019). African swine fever virus. In: Zimmerman, J. J., Karriker, L. A., Ramirez, A., Schwartz, K. J., Stevenson, G. W., Zhang, J. (Eds) *Diseases of Swine*, 11th edition, pp. 443-452. Wiley, New York.

Santonja, G. G., Georgitzikis, K., Scalet, B. M., Montobbio, P., Roudier, S. and Sancho, L. D. (2017). *Best available Techniques (BAT) Reference Document for the Intensive Rearing of Poultry or Pigs*. Publications Office of the European Union, Luxembourg.

Sarignac, C., Signoret, J. P. and McGlone, J. J. (1997). Relation mere-jeune, comportement et performances en function du Systeme de logement et de l'environnement social. Sow and piglet performance and behavior in either intensive outdoor or indoor units with litters managed as individuals or as small social groups. *Journees Rech. Porcine en France* 29, 123-128.

Sasaki, Y., Ushijima, R. and Sueyoshi, M. (2015). Field study of hind limb claw lesions and claw measures in sows. *Animal Science Journal* 86(3), 351-357.

Scheidt, A. B., Cline, T. R., Clark, L. K., Mayrose, V. B., Van Alstine, W. G., Diekman, M. A. and Singleton, W. L. (1995). The effect of all-in-all-out growing-finishing on the health of pigs. *Journal of Swine Health and Production* 3(5), 202-205.

Schmied, C., Boivin, X., Scala, S. and Waiblinger, S. (2010). Effect of previous stroking on reactions to a veterinary procedure: behaviour and heart rate of dairy cows. *Interaction Studies* 11(3), 467-481.

Schmolke, S. A., Li, Y. Z. and Gonyou, H. W. (2003). Effect of group size on performance of growing-finishing pigs. *Journal of Animal Science* 81(4), 874-878.

Schrøder-Petersen, D. L. and Simonsen, H. B. (2001). Tail biting in pigs. *The Veterinary Journal* 162(3), 196-210.

Schulenburg, A., von der Meyer, K. and Dierks-Meyer, B. (1986). Orthopedic and microstructural studies on claw horn of fattening pigs in various types of housing. *Veterinary Medicine* 125, 130-132.

Sørensen, J. T. and Schrader, L. (2019). Labelling as a tool for improving animal welfare—the pig case. *Agriculture* 9(6), 123.

Spoolder, H. A. M., Edwards, S. A. and Corning, S. (1999). Effects of group size and feeder space allowance on welfare in finishing pigs. *Animal Science* 69(3), 481-489.

Stolba, A. and Wood-Gush, D. G. M. (1984). The identification of behavioural key features and their incorporation into a housing design for pigs. *Annales de Recherches Veterinaires. Annals of Veterinary Research* 15(2), 287-299.

Straw, B. E. and Bartlett, P. (2001). Flank or belly nosing in weaned pigs. *Journal of Swine Health and Production* 9(1), 19-23.

Stricklin, W. R. and Mench, J. A. (1987). Social organization. Farm Animal Behavior. E. O. Price (Ed.). *Veterinary Clinics of North America. Food Animal Practice* 3(2), 307-322.

Sutherland, M. A., Niekamp, S. R., Johnson, R. W., Van Alstine, W. G. and Salak-Johnson, J. L. (2007). Heat and social rank impact behavior and physiology of PRRS-virus-infected pigs. *Physiology and Behavior* 90(1), 73-81.

SVC (1996). Report on the welfare of laying hens. European Commission, the Scientific Veterinary Committee Animal Welfare Section. Available at: http://ec.uropa.eu/food/fs/sc/oldcomm4/out33_en.pdf.

SVC (1997). The welfare of intensively kept pigs. Report of the Scientific Veterinary Committee of the EU. Available at: http://ec.europa.eu/food/animal/welfare/farm/out17_en.pdf.

Taylor, N., Prescott, N., Perry, G., Potter, M., Le Sueur, C. L. and Wathes, C. (2006). Preference of growing pigs for illuminance. *Applied Animal Behaviour Science* 96(1-2), 19-31.

Thacker, E. L. and Minion, F. C. (2012). Mycoplasmosis. In: Zimmermann, J. J., Karriker, L. A., Ramirez, A., Schwartz, K. J. and Stevenson, G. W. (Eds), *Diseases of Swine*, pp. 779-798. WileyBlackwell, Ames, IA.

Thomas, L. L., Goodband, R. D., Woodworth, J. C., Tokach, M. D., DeRouchey, J. M. and Dritz, S. S. (2017). Effects of space allocation on finishing pig growth performance and carcass characteristics. *Translational Animal Science* 1(3), 351-357.

Tubbs, R. C. (1988). Lameness in sows: solving a preventable problem. *Veterinary Medicine*.

Turner, S. P., Allcroft, D. J. and Edwards, S. A. (2003). Housing pigs in large social groups: a review of implications for performance and other economic traits. *Livestock Production Science* 82(1), 39-51.

Turner, S. P. and Edwards, S. A. (2004). Housing immature domestic pigs in large social groups: implications for social organisation in a hierarchical society. *Applied Animal Behaviour Science* 87(3-4), 239-253.

Tuyttens, F. A. M. (2005). The importance of straw for pig and cattle welfare: a review. *Applied Animal Behaviour Science* 92(3), 261-282.

Urbain, B., Gustin, P., Prouvost, J. F. and Ansay, M. (1994). Quantitative assessment of aerial ammonia toxicity to the nasal mucosa by use of the nasal lavage method in pigs. *American Journal of Veterinary Research* 55(9), 1335-1340.

Velarde, A., Fàbrega, E., Blanco-Penedo, I. and Dalmau, A. (2015). Animal welfare towards sustainability in pork meat production. *Meat Science* 109, 13-17.

Verdon, M. and Rault, J. L. (2018). Aggression in group housed sows and fattening pigs. In: Spinka, M. (Ed.) *Advances in Pig Welfare*, pp. 235-260. Woodhead Publishing, Sawston, UK.

Verdon, M., Hansen, C. F., Rault, J. L., Jongman, E., Hansen, L. U., Plush, K. and Hemsworth, P. H. (2015). Effects of group housing on sow welfare: a review. *Journal of Animal Science* 93(5), 1999-2017.

Vermeer, H. M., De Greef, K. H. and Houwers, H. W. J. (2014). Space allowance and pen size affect welfare indicators and performance of growing pigs under Comfort Class conditions. *Livestock Science* 159, 79-86.

Vermeij, I., Enting, J. and Spoolder, H. A. M. (2009). *Effect of Slatted and Solid Floors and Permeability of Floors in Pig Houses on Environment, Animal Welfare and Health and Food Safety: A Review of Literature (No. 186)*. Animal Sciences Group.

Vitt, R., Weber, L., Zollitsch, W., Hörtenhuber, S. J., Baumgartner, J., Niebuhr, K., Piringer, M., Anders, I., Andre, K., Hennig-Pauka, I., Schönhart, M. and Schauberger, G. (2017). Modelled performance of energy saving air treatment devices to mitigate heat stress for confined livestock buildings in Central Europe. *Biosystems Engineering* 164, 85-97.

Waiblinger, S., Boivin, X., Pedersen, V., Tosi, M. V., Janczak, A. M., Visser, E. K. and Jones, R. B. (2006). Assessing the human–animal relationship in farmed species: a critical review. *Applied Animal Behaviour Science* 101(3-4), 185-242.

Wang, C., Chen, Y., Bi, Y., Zhao, P., Sun, H., Li, J., Liu, H., Zhang, R., Li, X. and Bao, J. (2020). Effects of long-term gentle handling on behavioral responses, production performance, and meat quality of pigs. *Animals: an Open Access Journal from MDPI* 10(2), 330.

Weary, D. M., Appleby, M. C. and Fraser, D. (1999). Responses of piglets to early separation from the sow. *Applied Animal Behaviour Science* 63(4), 289-300.

Weary, D. M., Huzzey, J. M. and Von Keyserlingk, M. A. G. (2009). Board-invited review: using behavior to predict and identify ill health in animals. *Journal of Animal Science* 87(2), 770-777.

Webb, N. G. (1984). Compressive stresses on, and the strength of the inner and outer digits of pigs' feet, and the implications for injury and floor design. *Journal of Agricultural Engineering Research* 30, 71-80.

Weng, R. C., Edwards, S. A. and English, P. R. (1998). Behaviour, social interactions and lesion scores of group-housed sows in relation to floor space allowance. *Applied Animal Behaviour Science* 59(4), 307-316.

Wolter, B. F., Ellis, M., Curtis, S. E., Augspurger, N. R., Hamilton, D. N., Parr, E. N. and Webel, D. M. (2001). Effect of group size on pig performance in a wean-to-finish production system. *Journal of Animal Science* 79(5), 1067-1073.

Worobec, E. K., Duncan, I. J. H. and Widowski, T. M. (1999). The effects of weaning at 7, 14 and 28 days on piglet behaviour. *Applied Animal Behaviour Science* 62(2-3), 173-182.

Ye, Z., Zhang, G., Seo, I.-H., Kai, P., Saha, C. K., Wang, C. and Li, B. (2009). Airflow characteristics at the surface of manure in a storage pit affected by ventilation rate, floor slat opening, and headspace height. *Biosystems Engineering* 104(1), 97-105.

Zonderland, J. J., Wolthuis-Fillerup, M., Van Reenen, C. G., Bracke, M. B., Kemp, B., Den Hartog, L. A. and Spoolder, H. A. (2008). Prevention and treatment of tail biting in weaned piglets. *Applied Animal Behaviour Science* 110(3-4), 269-281.

Chapter 2

Welfare of pigs during finishing

Jonathan Amory, Writtle University College, UK; and Nina Wainwright, British Pig Executive (BPEX), UK

1 Introduction

Finisher pigs, those of a post-weaning age kept for slaughter, make up the majority of the world pig population of approximately 1 billion (2014 - latest figures from FAOstat accessed on 19 October 2017). Although many of the countries in the world keep pigs in small groups that scavenge or are fed household waste, high-income countries (per capita GDP levels in excess of 30 000 USD) raise more than 95% of their pigs in intensive systems (Gilbert et al., 2015). In the European Union (EU), neonates (piglets) may be reared in extensive outdoor systems along with the sow, but keeping post-weaned pigs in systems other than indoor housing is rare (European Food Safety Authority, 2007). The intensive production system is characterised by higher animal density, larger farms, use of concentrated foods and control of the production environment, particularly temperature, humidity and lighting. For finishing pigs there are a range of production systems available, each with their own advantages and disadvantages; however, the majority will be reared in a partly or fully slatted floor system with restricted or no access to straw (Table 1).

The welfare of an animal has been defined by its ability to cope with its environment (Broom, 1986); a finishing pig in any production system is exposed to a range of challenges that threaten its well-being. These challenges

http://dx.doi.org/10.19103/AS.2020.0081.06

Table 1 Housing systems for EU finishing pigs* (adapted from (Hendriks et al., 1999)

	Without/restricted access to straw			With straw	
System	Partly slatted	Fully slatted	Solid concrete	Solid concrete	Deep litter
%	47	44	3	4	2

* Note the total population in this report was stated to be approximately 69 million animals.

or stressors can be either physiological or psychological in nature, and both the mental health and physical health of an animal are important to its welfare (Farm Animal Welfare Council, 2009). In addition, welfare scientists have more recently focused on the positive aspects of welfare that are considered important for a sentient animal. Particular risks to the welfare of finisher pigs relate to nutritional management, the physical environment and a lack of mental stimulation leading to such problems as tail-biting. This chapter discusses these concerns and current practical measures adopted in welfare assessment.

2 Nutrition management and welfare of finishing pigs

2.1 Gastric ulcers

A key objective of pig production is to let animals attain their slaughter weight at least cost as quickly as possible. Intensive systems often feed finely ground pelleted diets to increase the digestibility of nutrients and reduce feed wastage, but this can have important consequences on finisher pigs' welfare. Abattoir studies from major pig-producing countries have reported high prevalence of gastric ulcers varying from 19% to 79% (Robertson et al., 2002; van den Berg et al., 2005; Amory et al., 2006; Swaby and Gregory, 2012; Omotosho et al., 2016).

There are many risk factors associated with gastric ulceration and most commonly with intensified husbandry; however, the main risk factors are those concerned with the diet. Interruption in the diet or fasting are major factors inducing gastric ulcers in pigs (Davies et al., 1994; Lawrence et al., 1998), but causes may vary from disease (Dybkjær et al., 1998) to general environmental stressors, such as transport (Lawrence et al., 1998), large group size (Baeckstroem et al., 1988), low social ranking (Hessing et al., 1994), providing dam rather than river water (Robertson et al., 2002) and a lack of environmental enrichment (Ramis et al., 2005). A particularly important dietary risk factor is the small particle size often associated with feeding a pelleted diet.

A number of authors have reported increased gastric ulceration with reduced particle size in the ground cereal proportion of the diet given to pigs (Potkins et al., 1989; Wondra et al., 1995a,b; Millet et al., 2012a,b; Cappai et al., 2013); however, a recent study found that particle size or pelleting has no effect on stomach ulceration (Ball et al., 2015). Cappai et al. (2013) described a mechanism for the pathogenesis due to the increased fluidity of stomach contents, an associated increase in acidity in the proximal part of the stomach and the lack of secretory cells in the oesophageal mucosa predisposing that region to developing bleeding gastric ulcers. Infection due to *Helicobacter suis* may also have an important role in ulcer formation (Haesebrouck et al., 2009). It is important to note that because gastric ulcers cause increased mortality in finishing pigs and are considered detrimental to economic production, they are an unfortunate trade-off in the pursuit of increased feed efficiency. The increasing demand for grain for human and animal food can be met by searching for alternative feed ingredients, for example, high-fibre by-products from food and bio-fuel production that may increasingly replace single grain- and soybean-based diets, though finding suitable alternatives is not without challenges (Bakare et al., 2013; Zijlstra and Beltranena, 2013). Crude fibre has been reported to have an additional dietary fibre that may break up the more fluid digesta from fine diets, preventing the erosion of the pars oesophageal region (Potkins et al., 1989). A number of reports have highlighted the protective effect straw bedding, presumably consumed as part of natural rooting activities, appears to have on gastric ulcer development (Potkins et al., 1989; Amory et al., 2006; Scott et al., 2006).

An additional benefit of a high-fibre diet may be the increased post-prandial satiety of pigs due to bulkiness of the fibre source or its ability to absorb water. Satiety, defined as an absence of hunger (Yeates and Main, 2008), is an important animal welfare concern and is an opportunity to establish positive well-being in pig production. Using pig behaviour to identify satiety is complex as gastric distension associated with a bulky food might reduce feeding behaviour through abdominal discomfort whilst the animal is still motivated to feed, due to nutrient restriction. Day et al. (1996) investigated this through use of a second-order reinforcement schedule where the pig was instructed to press a paddle to indicate feeding motivation thus removing the potential confounder of further ingestion of food. They found that dried citrus pulp, a fibre source of high water-holding capacity, reduced feeding motivation, for up to three hours post-feeding compared to only one hour by a low-fibre diet of an equivalent level of nutrients. There have been few studies of finishing pigs since, with most research carried out in sows where food deprivation is a well-recognised concern. Leeuw et al. (2008) provided a detailed review of both sows and finishing pigs to ascertain the welfare benefits of both bulky and fermentable fibres, concluding that it is fermentable dietary fibre that provides the longer term postprandial satiety effects, reducing physical activity and appetitive behaviour for many hours after a meal.

2.2 Nutritional deficiencies and pig behaviour

Gross inadequacies in nutrition provision will have obvious effects on pig behaviour and consequently welfare for example, lack of food leads to hunger (in itself a welfare concern) and can lead to an increase in foraging, and greater competition and aggressive acts around any remaining food or when food supply is renewed, resulting in fighting lesions and injuries (Day et al., 1995). Similar effects will also be seen, usually at lower level, when there are less obvious issues for example, acute or chronic nutrient imbalance; pigs have shown that they can accurately select between different diets in order to obtain the correct balance of specific amino acids (Kyriazakis et al., 1991) - this could be reflected in time spent foraging if these are not correctly balanced in their provided diet.

Whilst there may be intrinsically apparent biological pathways between nutritional deficits and behaviour (e.g. hunger → food searching), individual nutrients can have less obvious effects - for example, specific amino acids play a role in neurotransmission through the blood-brain barrier and can therefore affect the animal's behaviour at this precise chemical/physiological level, for example, Popova (2006). Edwards (2006) stated that a changed protein metabolism may pathologically affect neurotransmitter balances in the central nervous system. It has been suggested that individual pigs performing tail-biting at obsessively high levels may have reduced protein uptake - this has several similarities with metabolic disorders found in humans. Deficiencies in metabolism of phenylalanine (an amino acid) in humans lead to the rare condition of phenylketonuria (PKU) (Moyle et al., 2007), a symptom of disruptive behaviours such as self-harm or tantrums (Brumm et al., 2010); individuals with PKU often have low birth weight (Lee et al., 2005).

In several species, it is suggested that gut discomfort leads to increased chewing behaviour which stimulates saliva production - the additional saliva is thought to have a role in reducing gut discomfort. Kubo et al. (2015) also suggested that a repetitive chewing action in itself may calm pigs. Early stages of crib-biting in horses can be alleviated by providing them with antacids (Mills and Macleod, 2002); chewing behaviour, such as tail or ear-biting, in pigs may similarly be directed at objects within the pen, but may also be directed at pen-mates. Pigs with conditions such as ileitis (porcine proliferative enteropathy, PPE) or gastric ulcers may also therefore be more likely to tail-bite. Almond and Bilkei (2006) showed that groups of pigs which had been vaccinated against the bacterium *Lawsonia intracellularis* (associated with ileitis) had lower rates of cannibalism-related culling (an alternative description of severe tail-biting) and mortality.

2.3 Salt as a welfare concern

Salt is mainly considered as a welfare concern because pigs are susceptible to salt poisoning when dietary levels are too high (0.4–0.5%) and pigs are not ingesting sufficient water, (The Pig Site, n.d.). In many species, animals will seek out additional salt when stressed (Denton et al., 1999), and this can be recreated by administration of ACTH (adrenocorticotrophic hormone – a hormone produced in response to stress). It has therefore been suggested that stress and salt-seeking behaviour in pigs may be observed as increased foraging and chewing behaviour, potentially directed at pen-mates' tails; however, ACTH injection in the pig does not trigger such a response (Jankevicius and Widowski, 2003), but this does not rule out that stress does cause this behaviour, only that ACTH may not be the appropriate model. Salt-deficient pigs do show increased rooting behaviour (Beattie and O'Connell, 2002). Anecdotally, increasing salt levels at critical tail-biting periods on individual farms has been reported as being successful in reducing tail-biting behaviour, suggesting that on these farms, at these key periods, salt-seeking behaviour may be occurring in response to stressors; therefore, increasing the amount of salt in gradual pace is recommended, rather than feeding pigs initially with higher salt content.

3 Physical and social environment and welfare of finishing pigs

3.1 Group size, stocking density and competition for resources

Aggression is a natural behaviour in pigs with a function of improving social position (Camerlink et al., 2014) and possibly chasing away strangers (Puppe, 1998). However, it has an impact on their welfare in a number of ways: increasing skin lesions with their associated pain and risk of infection (Turner et al., 2006), stress physiology (Escribano et al., 2015) and probably becoming a part of a wider pattern of social tension in finishing pigs (Camerlink et al., 2014).

Various aspects of pen management affect levels of aggression in pigs. Pigs are regrouped for various management reasons, such as reducing stocking density or making efficient use of pen spaces, and can be grouped into a finisher category right up to the few weeks before being slaughtered (Stookey and Gonyou, 1994). When previously unacquainted finisher pigs are put together in a common pen, they show agonistic behaviour (aggressive and submissive actions) that varies in intensity over a day and that increases in frequency and fight time compared with newly weaned pigs (Stukenborg et al., 2011). Larger group sizes are associated with poor growth post-weaning; however, this association is not found in pigs above a live weight of 69 kg (Turner et al., 2003). There is also little evidence of a relationship between group size and aggression (Turner et al., 2001; O'Connell et al., 2004; Schmolke et al., 2004) although pigs from large-size groups were less aggressive in encountering those from other groups (Turner et al., 2001).

Reducing space allowance below a critical value (Box 1) results in a decline in growth rate, indicating reduced well-being (Gonyou et al., 2006), and may still be more economical (Edwards et al., 1988; Vermeer et al., 2014) epitomising the tension between production and welfare. An interdisciplinary Dutch welfare initiative experimented with a 'ComfortClass' pen design, with separated rooting and dunging areas and a space allowance of up to 2.4 m² per pig based on the perceived space required to meet the needs of the animals (de Greef et al., 2011). This study concluded that this level resulted in a good quality of life for these finishing pigs, based on the absence of signs of poor welfare. A follow-up study of the ComfortClass pen found detrimental effects of increased stocking density, even at elevated levels, and increased group sizes in terms of welfare and growth (Vermeer et al., 2014); however, the performance benefits of more space/lower group sizes did not offset the increased housing costs.

Box 1 Space allowance and pig welfare

Floor allowance for pig production is often calculated by the following formula:

$A = k*BW(0.667)$

where A = floor space allowance (m²/BW), BW = body weight (kg) and k = space allowance coefficient

Gonyou et al. (2006) investigated the effects of stocking density on production efficiency from a range of published articles and concluded that crowding was detrimental to pig production (measured as average daily gain) in fully and partially slatted systems when the value of k was less than 0.035.

Competition for access to pen resources can be important for welfare. Increased competition for feeding spaces is associated with increased aggression (Spoolder et al., 1999) with lightweight animals particularly at risk of welfare impairment (Rasmussen et al., 2006). Competition for enrichment devices such as a tyre swing or block of straw can also increase fighting in finisher pigs (Schaefer et al., 1990; Bulens et al., 2016). Pigs seem to be less competitive to access drinking facility (Turner et al., 1999).

An article by Bernard Peet, on the well-known The PigSite, describes the importance of pen design in managing lying and dunging behaviour in order to reduce social tension through competition for resources or disturbance of

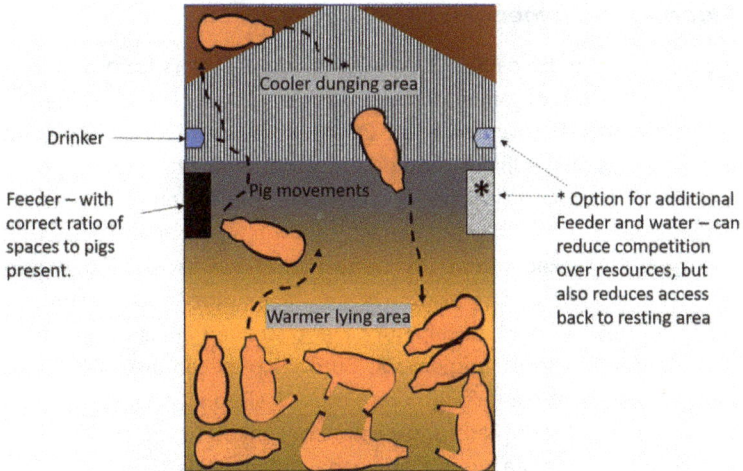

Figure 1 Optimum pen design to allow free movement and undisturbed lying behaviour (adapted from (Peet, n.d.)).

lying pigs (Peet, n.d.). Managing ventilation to provide a cooler dunging area, organising drinkers to be away from dunging areas, ensuring the lying area is as undisturbed as possible through correct positioning of feeders (Fig. 1) and ensuring pigs are not forced to lie in 'undesirable' areas will reduce aggression and possible tail-biting, as discussed in the following section.

3.2 Mounting behaviour

With surgical castration to reduce boar taint considered to be a major welfare concern, many major pig-producing countries in the EU have signed a common European declaration on alternatives. A 2014 progress report on this initiative describes one alternative as the rearing of entire males, commonly carried out in the United Kingdom, Spain and Portugal (Backus et al., 2014). However, this practice is associated with sexual behaviour manifesting as mounting of other animals that can result in skin lesions and leg injuries (Rydhmer et al., 2006; Fredriksen et al., 2008) that is a particular welfare concern to finisher pigs. Mounting is carried out mainly by male pigs mostly, but the behaviour is equally displayed towards male and female animals in mixed groups (Hintze et al., 2013). Immunocastration is not straightforward at present, but may provide a solution (Rydhmer et al., 2010); however, mounting behaviour can occur well before the onset of puberty (Clark and D'Eath, 2013) and is a trait that varies from pig to pig (Hintze et al., 2013). The underlying biological mechanisms and the approaches to reduce this behaviour within the industry are not clear at present.

3.3 Flooring and lameness

Lameness in pigs can be caused by a wide range of factors for example, injury or infection in the foot or joint, or longer term skeletal and joint problems such as osteochondrosis (Stavrakakis et al., 2014). Injury or infection (e.g. bacterial infection via abrasions) is likely to be due to suboptimal flooring (rough-edged flooring or inappropriate slot size or slat shape, permanently damp flooring) either during the finisher stage or carried through from earlier housing; it can also be caused or exacerbated by behaviours such as running, mounting, fighting or fleeing (which can also have genetic components). Steps within a pen can contribute to mechanical injury. Deep strawed areas can increase the risk of lameness due to entanglement; outdoor pigs can also risk damaging limbs while trying to move through deep mud or slipping on ice.

Lameness and flooring are unquestionably linked to pig production, and the main flooring/substrate types all have their inherent risks. The material used for making a floor needs to be nonabrasive, non-slippery and preferably elastic (to some extent) to cushion the pigs while moving or resting. Slatted flooring is more complex, with the construction, profile and size of slots (gaps) being important in protecting the feet from injury (European Food Safety Authority (EFSA), 2007). EFSA (2005) concluded that width of a gap should not exceed half the width of the contact area between the foot and the floor and the solid area between the gaps should be sufficient enough to support the foot; they also noted that whilst several studies produced recommendations of slat width based on specific weights of pig, these may be confounded by slat type, shape and material. Construction profile is critical: sharp edges may cause cut injuries as well as a compressive stress (Webb and Nilsson, 1983; Webb, 1984). Foot and leg injuries tend to be higher in pigs on slatted systems than those on solid floors (European Food Safety Authority, 2005), whereas pigs housed in unbedded systems have a greater severity of pressure and bursae than those in bedded systems. Some of these may not feel the pain because they have no lesion or infection, but all have a potential to have their locomotion and social interactions affected and hence welfare.

As noted in EFSA (2005) and in the development of Real Welfare protocols, observation of lesions and bursae on pigs' limbs can be particularly difficult in typical housing – this may mean that suitability of housing is suboptimal but goes unnoticed until lameness develops. It is difficult to get to eye level with pigs' legs within a pen, as observers are likely to see either inquisitive heads and backs, or the back of rear legs if the pigs are flighty. Flighty animals are at increased risk of leg injuries and lameness (Stavrakakis, 2014).

There is a genetic component to lameness in pigs, with leg weakness and osteochondrosis found to be genetically correlated with lean meat content, and this points to a negative effect of selecting for lean meat content in the carcase (European Food Safety Authority, 2007). Stavrakakis et al. (2014) used automated gait tracking of pigs in order to identify issues, with the aim of aiding farmers in selecting breeding animals.

Lameness is considered to be a key outcome measure or indicator of the welfare of pigs in individual, group and herd levels (alongside tail-biting and other measures); a high level of lameness can be due to a range of factors. Identifying and addressing these factors can therefore improve the welfare of current and future batches of pigs on the farm.

3.4 Outdoor production

Whilst outdoor housing of pigs is often seen by consumers as achieving the highest welfare standards, it has inherent welfare risks to manage, which are rarely found in indoor systems. Exposure to a wide range of wildlife increases the pigs' exposure to parasites (European Food Safety Authority, 2007); lack of access to wallows or shaded areas in hot temperatures can result in sunburn (NADIS, 2003). Excess rain can make the fields impassable to pigs, and frozen water supply and frozen ground are also some of the challenges to pigs.

Indoor systems (often termed 'intensive' but covering a wide range of housing and system types, population densities etc.) were developed commercially from the 1930s (Woods, 2012) in order to improve pig production by taking away stressors and production challenges associated with outdoor rearing. Pigs housed indoors can be kept at an optimal temperature range, with ready access to dry bedding, clean food and water, and pen layouts enabling the stockperson to observe the animals for problems more frequently and more accurately (mud coverage on outdoor animals, especially in summer months can make observation difficult). Indoor housing does take away the range of environmental stimulation that the pig receives, for example, temperature range, weather and light levels. This less varied environment can potentially lead animals that are less able to cope with changes in daily routine to activities such as movement around the unit or loading.

4 Environmental enrichment and tail-biting

4.1 Environmental enrichment

Effective environmental enrichment for animals is achieved by providing them with what meets their behavioural or psychological needs in a situation where food, water and social and physical comfort are already provided. For most animals in captivity this relates to the appetitive stage of food (or other resource) attainment because food presentation and availability

Figure 2 'Suitable rootables' – cartoon produced for the Tail Biting Project (2006-09): implementation of existing knowledge on tail-biting by the development and evaluation of prevention and outbreak husbandry advisory tools. Research partners: University of Bristol, University of Newcastle. Industrial partners: RSPCA, BPEX, pig producers.

are predictable (Jensen, 2002). Pigs have evolved to root and investigate their environment with their snout and specialised rostral disc to obtain a balanced and nutritional and satiating diet (down to specific amino acids); this behaviour can account for 10-20% of their active time (Stolba and Wood-Gush, 1989). Suitable environmental enrichment for pigs is therefore one that they can root, chew and ideally ingest with some nutritional benefits; where such substrates or objects are provided, pigs show less unwanted agonistic behaviours such as tail-biting and aggression/fighting. Finding the ideal environmental enrichment for pigs has been the subject of many research papers as this is an area of commercial and welfare significance (Fig. 2).

Whilst papers often focus on comparing a small number of different substrates or objects to find pigs' preference and effects on both desirable and undesirable behaviours, a different approach was followed by van de Weerd and Day (2009) who described objects based on their properties, rather than by size, substance, etc. This system makes comparison of enrichment types far more useful in a practical environment, as well as comparisons of research papers, and helps to avoid creating a prescriptive list of recommended objects or substrates. Pigs were shown to have long-term (five-day) preference for objects that were edible and destructible. Additional work suggests that nutritional objects are also preferred (Day et al., 1996). Studnitz et al. (2007) confirmed this with additional detail, finding that the best materials for maintaining pigs' investigatory behaviour are complex, changeable, destructible, manipulable and contain sparsely distributed edible parts. Grandin (1989) showed that pigs prefer clean objects; this can be an issue where objects have become detached from their suspending chain or rope and are rooted into the dunging area of the pen, if objects are suspended at pig-height near the dunging area of the pen, or in a crowded pen with no specific dunging area. Longevity of use of enrichment is a key factor in commercial units where objects cannot be replaced daily.

Commercial pigs experience a wide range of enrichment types, both intended and incidental. In practice, this means that pigs may be finding objects around the pen that they choose to root and chew on, but which have not been intentionally provided as enrichment, for example, wooden gates; pen fittings such as hinges, rubber curtains across kennel entrances; baler twine; wellington boots; etc. In outdoor systems turf, earth, wildlife, etc. might all be considered by the pigs as suitable rootable objects/substrates and be a satisfactory outlet for rooting and chewing behaviours. This distinction between intended and incidental enrichment only becomes an issue on-farm where pigs are damaging pen fittings, etc. resulting in increased farm maintenance time and costs, or where pen fittings are chewed into a dangerous state (revealing wires, sharp edges etc., or allowing pigs to mix). Whilst car tyres have historically been recommended as enrichment for pigs, their destructive capabilities meant that there is a welfare risk from the exposed wire core of the tyres, resulting in entanglement, distress and lesions (NADIS, 2002). In academic exercises or where enrichment-related welfare issues are being addressed, provision needs to be made to recognise all objects or substrates that the pigs are choosing to root or chew on, whether intentional or not. An alternative interpretation of the EU wording on provision of enrichment has been to state that enrichment is appropriate (i.e. sufficient quality and quantity) *only* when pigs are not tail-biting (Compassion in World Farming, 2011), but this overlooks the importance of many other factors in tail-biting – good enrichment can ameliorate some, but not all, stressors attributable to suboptimal husbandry (Edwards, 2006). Typical enrichment provision for different systems is shown in Table 2.

Six per cent of pigs in Europe are housed in straw pens, (the United Kingdom and Switzerland are atypical with 25% and 20%, respectively, of finisher pigs housed on straw) (European Food Safety Authority, 2007). Apart from fulfilling the key properties of enrichment, straw offers additional benefits of providing thermal and physical comfort (Fraser et al., 1991; Arey and Bruce, 1993; Lyons et al., 1995; Schrøder-Petersen and Simonsen, 2001) and gastric comfort: pigs that had straw bedding or received rolled cereals had nearly no gastric lesions (Nielsen and Ingvartsen, 2000) which may reduce tail-biting risk by reducing gut discomfort and drive to chew. Straw is also generally provided in much larger quantities than other forms of enrichment, increasing pigs' opportunity to interact with it and reducing competition among pigs to access it compared with objects (see Box 2 for key points).

Table 2 Typical enrichment provision for different systems of pig production

System	Enrichment provided	Enrichment properties	Notes
Indoor, slatted	Suspended objects, generally plastics or wood (branch segments) suspended from chain or rope. Floor-based objects such as large plastic containers, logs. Large stones or rounded pebbles. Specific objects such as acetyl pipe helicopters, commercially available objects such as Bite-Rite. Occasionally cardboard boxes	Wide range of enrichment properties, for example, large solid metal or plastic objects limited malleable or deformable properties; wood good difficult to keep clean; limited nutritional properties	Can be limited by cost per pen (large number of pens per farm) and replacement rate. Objects and broken objects must not risk injury to pigs (or farm staff) or disruption to mechanised slurry systems
Indoor, solid (or part solid) floor	As above plus opportunity for bedding substrate, generally straw, occasionally sawdust or shavings		Where slats are present, bedding provision is limited due to risk of blockage to slurry system
Solid floor	Straw provision in some or all of pen; depending on pen size, entire bales may be provided		Where straw is present as bedding, additional enrichment objects are provided less often
Deep straw system	Straw		
Outdoor-housing + paddock	Strawed lying area		Parasite risk from external contamination

Box 2 Straw – the key points

Barley straw preferred to wheat straw

Clean, dry, bright straw (rather than damp and mouldy or old, dry and dusty) (EFSA Panel on Animal Health and Welfare, 2014)

Long straw, rather than chopped (Pedersen et al., 2005)

Regular replenishment required, for example, daily (EFSA Panel on Animal Health and Welfare, 2014)

Also effective when provided solely as enrichment (dispenser) rather than bedding (EFSA Panel on Animal Health and Welfare, 2014)

Deep straw systems – lying area needs to be dry (DEFRA, 2003)

4.2 Removing straw

In many units pigs move between housing types as they mature, for example, from weaner housing to grower pens then finisher housing. Each housing type may have different enrichment (or straw) opportunities, for example, fully slatted weaner decks → strawed grower yards → slatted finisher pens. Pigs may also be moved between units at the different stages; some pigs may therefore experience both strawed and non-strawed pens. Day et al., (2002) suggested that a change in straw provision may be more important than the straw *per se* in altering pigs' behaviour for example, increasing tail-biting, as suggested by other authors (Ruiterkamp, 1985; Statham et al., 2011) and has also been found to be the case for enrichment removal (Munsterhjelm et al., 2009).

4.3 Tail-biting

Tail-biting is a multifactorial behaviour in pigs and can result in 'mild' lesions (e.g. underlying bruising but no visible surface lesions) through to severe damage (also described as cannibalism), and can affect a significant percentage of pigs on a unit. Tail-biting has both welfare and financial impacts on a unit, with severe outbreaks potentially affecting all or a majority of pigs on a farm. Harley et al. (2012) found that 58.1% of finisher pigs reaching slaughter had detectable tail lesions, with an additional 1.03% having severe lesions (based on 36 963 pigs, 99% of which were tail docked). As a consequence, 42% of profit margin was eroded at slaughter due to tail-biting with losses due to reduced growth and pig meat lost to carcase condemnation totalling €1.70 per pig (Harley et al., 2012). Despite routine tail docking (a measure used to reduce tail-biting risk) the overall cost of tail-biting to the Dutch pig industry was calculated to be €8 million each year (EFSA Panel on Animal Health and Welfare, 2014). Whilst tail-biting research has previously been considered in terms of the outcome of the behaviour and its obvious production and welfare impact that is, bitten tails, abscesses, potential pyaemia in the pig, on-farm deaths, partial and whole carcase condemnations, research has more recently focused on the welfare of the pigs driven to perform this behaviour, and therefore the potential to help prevent the behaviour, by either identifying individuals predisposed to tail-biting (Brunberg et al., 2013) or adjusting the environment to better adapt the pigs to. Pigs which tail-bite are demonstrating an abnormal behaviour (not reported in wild boar or feral pig populations) indicating an inability to cope with their environment.

Whilst tail-biting can occur at any stage of pig production, it has more obvious production and welfare impacts at finisher stages of production due to the size of lesion inflicted and the value of the animals at this stage.

Taylor et al. (2010) described three types of tail-biting based on the pigs' underlying motivation. The most commonly described tail-biting is boredom tail-biting, attributed to the pig being unable to find suitable rooting and chewing objects within its environment, resulting in rooting and chewing directed towards the tail (and/or ears, especially if the tail is docked to a short length); this behaviour is thought to start with innocuous manipulation of the tail in the mouth (described as 'tail in mouth' behaviour by Schrøder-Petersen and Simonsen, 2001), and to develop biting and chewing in some cases into stronger biting and chewing, resulting in lesions. A second type of tail-biting is frustration tail-biting, where a pig bites the tail of another pig that is blocking its access to resources such as feed, transit through an area of the pen, or to optimal lying areas; the biter pig may learn that this behaviour is rewarded with access to what it wants, and so the pattern is reinforced. A third category of tail-biting is generally observed in fanatic pigs, often the smallest in a group, which perform tail-biting at unusual (and sometimes extremely high) levels; the underlying cause of this behaviour in these individuals is unknown, but it may be linked to their slower growth for example, due to protein imbalance (and the diet provided to pigs in their group may not cater to their individual needs). By identifying the underlying cause of tail-biting, it is expected that tail-biting can be prevented or reduced by addressing the specific problem for that group, building or unit. This approach does not rule out the possibility of other tail-biting ontogenies such as tail suckling in weaners. Once tail-biting occurs within a group, other pigs of the same group may become attracted to the scent of blood (and potentially the protein or salt content of the blood) and can result in outbreak behaviour where significant numbers of pigs become involved, both as biters and victims (see Box 3 for Genetics and Tail-biting).

Box 3 Genetics and tail-biting

Not all pigs perform tail-biting under the same environment and stressors as their pen-mates; therefore, internal variation in propensity is also a factor in tail-biting. This genetic component is also the subject of study (e.g. Breuer et al., 2005) as identifying and removing high-risk factors from the commercial pig population would help with pig management and improve the welfare of the commercial pig herd (provided tail-biting propensity is not linked to beneficial behavioural or disease-susceptible traits). There are correlations among tail-biting, lean tissue growth rate and back fat thickness (Beattie et al., 2005); lower back fat thickness also emerged as a risk factor (Moinard et al., 2003). However, lean meat is perceived as a desirable product by consumers, and Breuer et al., (2005) suggested that selection for a leaner carcase in modern pig production has resulted in pigs that are more prone to tail-biting.

Boredom tail-biting is likely to be best addressed by providing optimal environmental enrichment such as straw or other suitable substrates or objects meeting the criteria established by Van de Weerd et al. (2003). Frustration tail-biting should be reduced by identifying the critical resource or resources and by increasing access to them, for example, by adding feed spaces, removing cinch points in traffic around the pen and altering pen layout to stop traffic across intended lying areas. Fanatic pigs are best dealt with by identifying the individual pigs that are tail-biting and removing them from the pen. Housing pigs with a range of tail lengths in the same group is a significant factor in tail-biting for example, housing short and long docked pigs together, undocked and half-docked pigs (reviewed by Taylor et al., 2012). Whilst variation in appearance of tail lengths is likely to be a factor in higher tail-biting due to novel tail shape being of more interest, a range of background factors may also be important (Box 4).

Box 4 Variation in tail lengths

Mixed groups - presence of two tail lengths within a group can indicate mixing of two or more litters (handled by two or more stockpeople when tail docked) - mixing of pigs is generally avoided due to associated aggression and stressors.

'Sunday pigs' - piglets born at the weekend - may not receive all procedures that other litters receive as well as tail docking; this could include procedures such as iron injections, with subsequent effects on health.

Gilts - female pigs identified as potential future breeding animals - may be left long-tailed to help identify them at sorting (short docked tails are avoided for breeding gilts as the projecting stump is considered to make mounting uncomfortable for the boar). Gilts present in a finisher group have therefore been remixed with mainstream finishers, with attendant mixing stressors.

Runts - may not be tail docked in some instances (to avoid additional stress to a weak piglet) - when present in mainstream groups may therefore be targeted as weaker individuals, but are also potentially more likely to become tail-biters.

Variation in tail lengths can be avoided by

1   Setting a standard tail length on the unit (based on length of tail remaining, not length removed)

2   Streamlining pig movements on the farm so that pig groups are not mixed later in the system (e.g. permanent gilt groups or 'smalls'/runt groups, or do not regroup different tail length groups if stockpeople cannot agree on a set length to dock to)

3   Enabling pigs to be trained and be familiar with docking equipment to reduce variation by individual stockpeople (and review of all other training to ensure no other aspects of pig keeping are being overlooked)

5 Practical welfare assessment of finisher pigs

5.1 Outcomes approach to welfare assessment

Welfare assessments have generally been based on physically measurable features of the animals' environment ('inputs') for example, pen size, temperature, water flow rate, etc. but this approach has been challenged in recent years by the 'Outcomes' approach. Outcome measures (or welfare outcomes) are based on observations of the animals for example, their behaviour, presence and location of lesions, lameness, etc. Outcome measures have been developed, validated and refined for pigs by programmes such as Welfare Quality (European-wide development, initiated in 2004), Real Welfare (initial work by S. Mullan (Mullan, 2009)) currently applied to all Red Tractor finisher units - UK) and AssureWel (RSPCA/Soil Association - applied to their certified finisher units).

Outcome schemes collect information from a representative sample of animals per unit (or per scheme) depending on the specificity of information needed, and what will be done with the information for example, a small sample from a large number of units to give a scheme-wide or national overview, or a large enough sample per unit to be able to benchmark and compare specific welfare measures accurately between different units and measure changes over time at individual unit level (Mullan et al., 2009). The sample pigs are examined by the assessor using specific criteria for each outcome measure, and presence or absence of the welfare measure is recorded (or levels of severity depending on the measure and scheme – see additional material). A key factor in valid use of outcome measures is adhering to the definitions to make objective rather than subjective assessments, and ensuring that assessors have been standardised (Mullan et al., 2011) – see Box 5.

Box 5 Definitions of lameness in welfare assessment

All three major outcome-based welfare assessment schemes recognise the importance of lameness as an indicator of welfare and management on a unit. The definition as used in Real Welfare has been agreed by vets to describe a level of lameness of significant welfare concern with affected pigs experiencing pain and/or discomfort:

Record as lame: any pig that when standing will not bear full weight on the affected limb and/or appears to be standing on its toes. When moving, there is a shortened stride with minimum or no weight-bearing on the affected limb and a swagger of the hind quarters. The pig may still be able to trot and gallop.

Additionally, consider whether pig should also be recorded under hospital pig category. This protocol does not include pigs that are showing stiffness or uneven gait as this is more subjective and difficult to interpret (AHDB Pork).

Identifying that an outcome such as lameness is 'an issue' (i.e. abnormally high) on a unit is just the starting point – the aim of these programmes is not just to record the prevalence of each measure, but to investigate and address likely causes of the problem.

5.2 Play as a positive welfare indicator

In addition to looking for indicators of negative or poor welfare in pigs, it is possible to look for positive behaviours such as play as well (Lawrence, 1987). Play in pigs can be locomotor or object based, and can be social or isolated (Newberry et al., 1988). Careful observation is needed to separate object-based play from investigatory behaviour and social/locomotor behaviour from fighting. Play behaviour can include throwing objects in the air and holding an object whilst being chased, rather than nudging, rooting and biting which would be more investigatory. Social play in pigs can also be difficult to distinguish from fighting, for example, behaviours such as 'shove' may be difficult to classify in all situations (Newberry et al., 1988), and it is likely that behaviour which began as play may turn into more agonistic behaviour for example, if playing pigs disturb resting animals. Play behaviour in pigs can be accompanied by a range of identifiable vocalisations which help to characterise the behaviours (Reimert et al., 2013) – where observations do not include sound it will be extremely difficult to identify all occasions of play. Social play in pigs can be instigated by head bobs, similar to play bows in dogs (Newberry et al., 1988). Studies on play in pigs by Wright (2012) concluded that the absence of play behaviour in indoor weaner and grower/finisher pigs could be evidence of poor welfare, (and presence of play behaviour in adult sows could be viewed as positive welfare) due to the prevalence of play behaviours in the different age groups (juvenile pigs in indoor housing would be expected to be observed playing in a functional observation window). As often noted, the presence of an observer at the pen will affect pigs' behaviour (Martin and Bateson, 2007). Projects comparing welfare outcome measures come up across several obstacles when trying to incorporate play as a welfare indicator, and it is generally not included in final protocols. One difficulty is in forming a clear definition of the behaviour that can be reliably recorded by numerous observers, given the subtle differences in between play and non-play, as well as practical difficulties in observing pigs within a commercial environment. The potentially short duration of the behaviour in finishers also meant that observation could be hit and miss. Presence of a novel observer could also affect pigs' likelihood of playing differentially in different environments.

5.3 *Vocalisations as a welfare indicator*

Vocalisation by pigs is now being recognised as a useful indicator of their behaviour and welfare, with automated systems being developed (e.g. Schön et al., 2004; Vandermeulen et al., 2015). High pitched vocalisations can be an indication of pain, distress or fear, for example, pigs being squashed due to huddling behaviour in low temperatures (Hillmann et al., 2004), competition and aggression when/if feed becomes limited or tail-biting incidence (Vandermeulen et al., 2015). Pig buildings without any stressors are likely to be characterised by low-pitched contentment grunts and few/no squeals. Friel et al. (2016) demonstrated that pigs from more stressed environments had lower rates of grunting than those from more enriched environments. Grandin (2010) used pig squeals when handled (at abattoir) as an indication of welfare, with plants passing if less than 5% of pigs squealed during the handling process (a series of potentially stressful events). Barking vocalisations are usually the result of pigs being startled ('alarm calls' – Chan et al., 2011) – a commercial environment should aim to reduce extreme startling events, for example, with handlers making pigs aware that they have entered a building, following set routines and introducing novel objects with care.

These different vocalisations can be readily picked up with experience, but can be difficult to characterise for use by unfamiliar observers. McLeman et al. (2008) showed that juvenile pigs (growers/finishers) could distinguish between conspecifics (both familiar and unfamiliar) by use of auditory cues alone. Provision of an acoustic environment which infringes on pigs' ability to hear and communicate can be assumed to reduce their welfare. The DEFRA code of regulations for pigs (DEFRA, 2003) states that 'Pigs shall not be exposed to constant or sudden noise. Noise levels above 85 dBA shall be avoided in that part of any building where pigs are kept'.

6 Future trends

Tear staining (chromodacryorrhea) is the occurrence of dark staining below the inner corner of the eye. Whilst this has been considered a reliable indicator of stress in rats (Harkness and Ridgway, 1980) and considered as a general indicator of welfare in pigs relating to air quality and clinical disease (Gruber, 2002; Whay et al., 2007), it has only recently been investigated as a specific measure of stress in pigs. Tear staining in pigs has now been shown to correlate with stress in pigs under controlled conditions, and more recently in pigs under commercial conditions: tear staining scores correlated with tail and ear damage, and lower tear staining scores were found in groups housed with environmental enrichment objects rather than control groups (Telkänranta et al., 2016). Presence of high degree of tear staining in a population could therefore be a useful starting point to investigate potential stressors.

7 Conclusion

There are particular challenges to the welfare of finishing pigs due to their significant physiological and behavioural needs combined with the economic pressures that drive intensive production. Global issues, particularly population growth combined with a growing demand for meat products from an increasingly urban population (Satterthwaite et al., 2010), are a potential threat to animal welfare with the associated drive towards higher levels of intensive production. Extensive reports of these issues have resulted in the inclusion of the welfare of finisher pigs in major programmes, such as Welfare Quality and the EU PiG innovation group (part of the EU Horizon 2020 programme). The World Organisation for Animal Health (OIE) first published a work on international standards on animal welfare in 2005, with 2016 seeing the 25th edition of the OIE Terrestrial Animal Health Code that makes general recommendations for a range of livestock species (OIE, 2016a) - a specific chapter on pig welfare is currently in development (OIE, 2016b). Recent international advances have been reported in regions such as Asia, the Far East (Murray et al., 2014) including the first Compassion in World Farming Good Pig Production awards launched in China in 2014 (Compassion in World Farming, n.d.).

Researchers working with industries have increasingly focused on practical welfare assessment and animal-based outcomes, as exemplified by Welfare Quality and the Real Welfare and AssureWel schemes in the United Kingdom. However, perhaps the greatest potential for further improvement comes from technological advancement. Precision Livestock Farming (PLF) is the automated continuous monitoring of animals. PLF approaches may be used in future to monitor a range of indicators pertinent to pig welfare including weight gain (Lu et al., 2016), drinking behaviour (Kashiha et al., 2013), aggression (Oczak et al., 2014), lying patterns (Nasirahmadi et al., 2015), lameness (Stavrakakis et al., 2015) and mounting behaviour (Nasirahmadi et al., 2016). Improvements in genetic technologies may provide solutions to particular finisher pig welfare solutions including breeding animals that are less aggressive (D'Eath et al., 2009), have better leg conformation to reduce lameness (Lu et al., 2016) and are less susceptible to heat stress (Rauw et al., 2017).

With an historic emphasis on suffering negative welfare indicators are well established, but increasing attention is being given to the concept of positive welfare (Mendl and Paul, 2004; Boissy et al., 2007; Mellor, 2012). In an influential 2009 report, the Farm Animal Welfare Council suggested that farm animals should have a 'Life Worth Living' and that the 'balance of an animal's experiences must be positive over its lifetime, (Farm Animal Welfare Council, 2009). Whilst this might be arguable (can negative experiences really be mitigated in this way?), clearly robust measures of both negative and positive welfare are needed to determine whether an animal does indeed have a life worth living. Future welfare codes are likely to include positive affective states as requirements alongside the minimisation of factors causing a reduction in well-being. Behaviours such as play and social interaction may provide useful indicators, but perhaps validation of positive indicators provides the main challenge in future pig welfare research.

8 Where to look for further information

Flooring and lameness:
    http://pork.ahdb.org.uk/media/2113/Action-37-Flooring.pdf4

Outcomes approach to welfare assessment:
    http://pork.ahdb.org.uk/health-welfare/welfare/real-welfare/
    http://www.welfarequality.net/network/45848/7/0/40
    http://www.assurewel.org/pigs

9 References

Almond, P. K. and G. Bilkei. 2006. Effects of oral vaccination against Lawsonia intracellularis on growing-finishing pig's performance in a pig production unit with endemic porcine proliferative enteropathy (PPE). *DTW Dtsch. Tierarztl. Wochenschr.* 113:232-5.

Amory, J. R., A. M. Mackenzie and G. P. Pearce. 2006. Factors in the housing environment of finisher pigs associated with the development of gastric ulcers. *Vet. Rec.* 158:260-4.

Arey, D. S. and J. M. Bruce. 1993. A note on the behaviour and performance of growing pigs provided with straw in a novel housing system. *Anim. Prod.* 56:269-72.

Backus, G., S. Støier, M. Courat, M. Bonneau and M. Higuera. 2014. First progress report from the European declaration on alternatives to surgical castration of pigs. Available from: https://ec.europa.eu/food/sites/food/files/animals/docs/aw_prac_farm_pigs_cast-alt_declaration_progress-report_20141028.pdf.

Baeckstroem, L., U. Wisconsin, T. Crenshaw and D. Shenkman. 1988. Gastric ulcers in swine: Effects of dietary fiber, corn particle size and preslaughter stress. In: *Rapport - Sveriges Lantbruksuniversitet, Veterinaermedicinska Fakulteten.* Institutionen foer Husdjurshygien, Sweden. Available from: http://agris.fao.org/agris-search/search.do?recordID=SE8811273.

Bakare, A. G., S. P. Ndou, and M. Chimonyo. 2013. Influence of physicochemical properties of fibrous diets on behavioural reactions of individually housed pigs. *Livest. Sci.* 157:527-34.

Ball, M. E. E., E. Magowan, K. J. McCracken, V. E. Beattie, R. Bradford, A. Thompson and F. J. Gordon. 2015. An investigation into the effect of dietary particle size and pelleting of diets for finishing pigs. *Livest. Sci.* 173:48-54.

Beattie, V. E. and N. E. O'Connell. 2002. Relationship between rooting behaviour and foraging in growing pigs. *Anim. Welf.* 11:295-303.

Beattie, V. E., K. Breuer, N. E. O'Connell, I. A. Sneddon, J. T. Mercer, K. A. Rance, M. E. M. Sutcliffe and S. A. Edwards. 2005. Factors identifying pigs predisposed to tail biting. *Anim. Sci.* 80:307-12.

Boissy, A., G. Manteuffel, M. B. Jensen, R. O. Moe, B. Spruijt, L. J. Keeling, C. Winckler, B. Forkman, I. Dimitrov, J. Langbein, M. Bakken, I. Veissier and A. Aubert. 2007. Assessment of positive emotions in animals to improve their welfare. *Physiol. Behav.* 92:375-97.

Breuer, K., M. E. M. Sutcliffe, J. T. Mercer, K. A. Rance, N. E. O'Connell, I. A. Sneddon and S. A. Edwards. 2005. Heritability of clinical tail-biting and its relation to performance traits. *Livest. Prod. Sci.* 93:87-94.

Broom, D. M. 1986. Indicators of poor welfare. *Br. Vet. J.* 142:524-6.

Brumm, V. L., D. Bilder and S. E. Waisbren. 2010. Psychiatric symptoms and disorders in phenylketonuria. *Mol. Genet. Metab.* 99(Suppl. 1):S59-63.

Brunberg, E., P. Jensen, A. Isaksson and L. J. Keeling. 2013. Behavioural and brain gene expression profiling in pigs during tail biting outbreaks – Evidence of a tail biting resistant phenotype. *PLoS ONE* 8:e66513.

Bulens, A., S. Van Beirendonck, J. Van Thielen, N. Buys and B. Driessen. 2016. Long-term effects of straw blocks in pens with finishing pigs and the interaction with boar type. *Appl. Anim. Behav. Sci.* 176:6-11.

Camerlink, I., S. P. Turner, W. W. Ursinus, I. Reimert and J. E. Bolhuis. 2014. Aggression and Affiliation during Social Conflict in Pigs. *PLoS ONE* 9:e113502.

Cappai, M. G., M. Picciau and W. Pinna. 2013. Ulcerogenic risk assessment of diets for pigs in relation to gastric lesion prevalence. *BMC Vet. Res.* 9:36. doi:10.1186/1746-6148-9-36.

Chan, W. Y., S. Cloutier and R. C. Newberry. 2011. Barking pigs: Differences in acoustic morphology predict juvenile responses to alarm calls. *Anim. Behav.* 82:767-74.

Clark, C. C. A. and R. B. D'Eath. 2013. Age over experience: Consistency of aggression and mounting behaviour in male and female pigs. *Appl. Anim. Behav. Sci.* 147:81-93.

Compassion in World Farming. 2011. Providing enrichment for pigs. Available from: https://www.ciwf.org.uk/media/3818880/providing-enrichment-for-pigs.pdf.

Compassion in World Farming. nd. China Awards | Compassion in Food Business. Available from: https://www.compassioninfoodbusiness.com/awards/china-awards/.

Davies, P. R., J. J. Grass, and W. E. (University of M. (USA) C. of V. M. Marsh. 1994. Time of slaughter affects prevalence of lesions of the pars oesophagea of pigs. Available from: http://agris.fao.org/agris-search/search.do?recordID=TH2000001484.

Day, J. E. L., I. Kyriazakis and A. B. Lawrence. 1995. The effect of food deprivation on the expression of foraging and exploratory behaviour in the growing pig. *Appl. Anim. Behav. Sci.* 42:193-206.

Day, J. E. L., I. Kyriazakis and A. B. Lawrence. 1996. The use of a second-order schedule to assess the effect of food bulk on the feeding motivation of growing pigs. *Anim. Sci.* 63:447-55.

Day, J. E. L., A. Burfoot, C. M. Docking, X. Whittaker, H. A. M. Spoolder and S. A. Edwards. 2002. The effects of prior experience of straw and the level of straw provision on the behaviour of growing pigs. *Appl. Anim. Behav. Sci.* 76:189-202.

D'Eath, R. B., R. Roehe, S. P. Turner, S. H. Ison, M. Farish, M. C. Jack and A. B. Lawrence. 2009. Genetics of animal temperament: Aggressive behaviour at mixing is genetically associated with the response to handling in pigs. *Animal* 3:1544-54.

DEFRA. 2003. *Code of Recommendations for the Welfare of Livestock: Pigs*. DEFRA Publications, London.

Denton, D. A., J. R. Blair-West, M. I. McBurnie, J. A. P. Miller, R. S. Weisinger and R. M. Williams. 1999. Effect of adrenocorticotrophic hormone on sodium appetite in mice. *Am. J. Physiol. - Regul. Integr. Comp. Physiol.* 277:R1033-40.

de Greef, K. H., H. M. Vermeer, H. W. J. Houwers and A. P. Bos. 2011. Proof of Principle of the Comfort Class concept in pigs: Experimenting in the midst of a stakeholder process on pig welfare. *Livest. Sci.* 139:172-85.

Dybkjær, L., L. G. Paisley, L. Vraa-Andersen and G. Christensen. 1998. Associations between behavioural indicators of `stress' in weaner pigs and respiratory lesions at slaughter. *Prev. Vet. Med.* 34:175-90.

Edwards, S. A. 2006. Tail biting in pigs: Understanding the intractable problem. *Vet. J.* 171:198-9.

Edwards, S. A., A. W. Armsby and H. H. Spechter. 1988. Effects of floor area allowance on performance of growing pigs kept on fully slatted floors. *Anim. Sci.* 46:453-9.

EFSA Panel on Animal Health and Welfare. 2014. Scientific Opinion concerning a Multifactorial approach on the use of animal and non-animal-based measures to assess the welfare of pigs. *EFSA J.* 12:3702.

Escribano, D., A. M. Gutiérrez, F. Tecles and J. J. Cerón. 2015. Changes in saliva biomarkers of stress and immunity in domestic pigs exposed to a psychosocial stressor. *Res. Vet. Sci.* 102:38-44.

European Food Safety Authority. 2007. Opinion of the Scientific Panel on Animal Health and Welfare on a request from the Commission related to animal health and welfare in fattening pigs in relation to housing and husbandry. *EFSA J.* 5:564.

European Food Safety Authority (EFSA). 2005. Opinion of the Scientific Panel on Animal Health and Welfare (AHAW) on a request from the Commission related to welfare of weaners and rearing pigs: Effects of different space allowances and floor. *EFSA J.* 3:585.

Farm Animal Welfare Council. 2009. Farm animal welfare in Great Britain: Past, present and future. Available from: https://www.gov.uk/government/uploads/system/uploads/attachment_data/file/319292/Farm_Animal_Welfare_in_Great_Britain_-_Past__Present_and_Future.pdf.

Fraser, D., P. A. Phillips, B. K. Thompson and T. Tennessen. 1991. Effect of straw on the behaviour of growing pigs. *Appl. Anim. Behav. Sci.* 30:307-18.

Fredriksen, B., B. M. Lium, C. H. Marka, B. Mosveen and O. Nafstad. 2008. Entire male pigs in farrow-to-finish pens–Effects on animal welfare. *Appl. Anim. Behav. Sci.* 110:258-68.

Friel, M., H. P. Kunc, K. Griffin, L. Asher and L. M. Collins. 2016. Acoustic signalling reflects personality in a social mammal. *R. Soc. Open Sci.* 3:160178.

Gilbert, M., G. Conchedda, T. P. V. Boeckel, G. Cinardi, C. Linard, G. Nicolas, W. Thanapongtharm, L. D'Aietti, W. Wint, S. H. Newman and T. P. Robinson. 2015. Income disparities and the global distribution of intensively farmed chicken and pigs. *PLoS ONE* 10:e0133381.

Gonyou, H. W., M. C. Brumm, E. Bush, J. Deen, S. A. Edwards, T. Fangman, J. J. McGlone, M. Meunier-Salaun, R. B. Morrison, H. Spoolder, P. L. Sundberg and A. K. Johnson. 2006. Application of broken-line analysis to assess floor space requirements of nursery and grower-finisher pigs expressed on an allometric basis. *J. Anim. Sci.* 84:229-35.

Grandin, T. 1989. Effect of rearing environment and environmental enrichment on behavior and neural development in young pigs. Available from: http://agris.fao. org/agris-search/search.do?recordID=US201300166011.

Grandin, T. 2010. Auditing animal welfare at slaughter plants. *Meat Sci.* 86:56-65.

Gruber, T. 2002. Aufstallung, Fütterung, Hygiene, Gesundheit und Management von Mastschweinen in biologisch bewirtschafteten Betrieben. PhD thesis, University of Veterinary Medicine, Vienna.

Haesebrouck, F., F. Pasmans, B. Flahou, K. Chiers, M. Baele, T. Meyns, A. Decostere and R. Ducatelle. 2009. Gastric helicobacters in domestic animals and nonhuman primates and their significance for human health. *Clin. Microbiol. Rev.* 22:202-23.

Harkness, J. and M. Ridgway. 1980. Chromodacryorrhea in laboratory rats (Rattus norvegicus): Etiologic considerations. *Lab. Anim. Sci.* 30:841-4.

Harley, S., S. J. More, N. E. O'Connell, A. Hanlon, D. Teixeira and L. Boyle. 2012. Evaluating the prevalence of tail biting and carcase condemnations in slaughter pigs in the Republic and Northern Ireland, and the potential of abattoir meat inspection as a welfare surveillance tool. *Vet. Rec.* 171:621.

Hendriks, H. J. M., A. M. Van de Weerdhof and N. InfoMil. 1999. Dutch notes on BAT for pig and poultry intensive livestock farming. Minist. Hous. Spat. Plan. Environ. MANMF. Available from: http://files.gamta.lt/aaa/Tipk/tipk/4_kiti%20GPGB/37.pdf.

Hessing, M. J. C., A. M. Hagelsø, W. G. P. Schouten, P. R. Wiepkema and J. A. M. Van Beek. 1994. Individual behavioral and physiological strategies in pigs. *Physiol. Behav.* 55:39-46.

Hillmann, E., C. Mayer, P.-C. Schön, B. Puppe and L. Schrader. 2004. Vocalisation of domestic pigs (Sus scrofa domestica) as an indicator for their adaptation towards ambient temperatures. *ResearchGate* 89:195-206.

Hintze, S., D. Scott, S. Turner, S. L. Meddle and R. B. D'Eath. 2013. Mounting behaviour in finishing pigs: Stable individual differences are not due to dominance or stage of sexual development. *Appl. Anim. Behav. Sci.* 147:69-80.

Jankevicius, M. L. and T. M. Widowski. 2003. Exogenous adrenocorticotrophic hormone does not elicit a salt appetite in growing pigs. *Physiol. Behav.* 78:277-84. doi:10.1016/ S0031-9384(02)00970-8.

Jensen, P. 2002. *The Ethology of Domestic Animals: An Introductory Text.* CABI Pub, Wallingford.

Kashiha, M., C. Bahr, S. A. Haredasht, S. Ott, C. P. H. Moons, T. A. Niewold, F. O. Ödberg and D. Berckmans. 2013. The automatic monitoring of pigs water use by cameras. *Comput. Electron. Agric.* 90:164-9.

Kubo, K., M. Iinuma and H. Chen. 2015. Mastication as a stress-coping behavior. *BioMed Res. Int.* 2015:11. doi:10.1155/2015/876409. Available from: http://www.ncbi.nlm.nih.gov/ pmc/articles/PMC4450283/.

Kyriazakis, I., G. C. Emmans and C. T. Whittemore. 1991. The ability of pigs to control their protein intake when fed in three different ways. *Physiol. Behav.* 50:1197-203.

Lawrence, A. 1987. Consumer demand theory and the assessment of animal welfare. *ResearchGate* 35:293-5.

Lawrence, B. V., D. B. Anderson, O. Adeola and T. R. Cline. 1998. Changes in pars esophageal tissue appearance of the porcine stomach in response to transportation, feed deprivation, and diet composition. *J. Anim. Sci.* 76:788-95.

Lee, P. J., D. Ridout, J. H. Walter and F. Cockburn. 2005. Maternal phenylketonuria: Report from the United Kingdom Registry 1978-97. *Arch. Dis. Child.* 90:143-6.

Leeuw, J. A. de, J. E. Bolhuis, G. Bosch and W. J. J. Gerrits. 2008. Effects of dietary fibre on behaviour and satiety in pigs: Symposium on 'Behavioural nutrition and energy balance in the young'. *Proc. Nutr. Soc.* 67:334-42.

Lu, M., Y. Xiong, K. Li, L. Liu, L. Yan, Y. Ding, X. Lin, X. Yang and M. Shen. 2016. An automatic splitting method for the adhesive piglets' gray scale image based on the ellipse shape feature. *Comput. Electron. Agric.* 120:53-62.

Lyons, C. A. P., J. M. Bruce, V. R. Fowler and P. R. English. 1995. A comparison of productivity and welfare of growing pigs in four intensive systems. *Livest. Prod. Sci.* 43:265-74.

Martin, P. and P. Bateson. 2007. *Measuring Behaviour: An Introductory Guide*. Updated edition edition. Cambridge University Press, Cambridge and New York.

McLeman, M. A., M. T. Mendl, R. B. Jones and C. M. Wathes. 2008. Social discrimination of familiar conspecifics by juvenile pigs, Sus scrofa: Development of a non-invasive method to study the transmission of unimodal and bimodal cues between live stimuli. *Appl. Anim. Behav. Sci.* 115:123-37.

Mellor, D. J. 2012. Animal emotions, behaviour and the promotion of positive welfare states. *N. Z. Vet. J.* 60:1-8.

Mendl, M. and E. Paul. 2004. Consciousness, emotion and animal welfare: Insights from cognitive science. *Anim. Welf.* 13:17-25.

Millet, S., S. Kumar, J. De Boever, R. Ducatelle and D. De Brabander. 2012a. Effect of feed processing on growth performance and gastric mucosa integrity in pigs from weaning until slaughter. *Anim. Feed Sci. Technol.* 175:175-81.

Millet, S., S. Kumar, J. De Boever, T. Meyns, M. Aluwé, D. De Brabander and R. Ducatelle. 2012b. Effect of particle size distribution and dietary crude fibre content on growth performance and gastric mucosa integrity of growing-finishing pigs. *Vet. J.* 192:316-21.

Mills, D. S. and C. A. Macleod. 2002. The response of crib-biting and windsucking in horses to dietary supplementation with an antacid mixture. *Ippologia* 13:33-41.

Moinard, C., M. Mendl, C. J. Nicol and L. E. Green. 2003. A case control study of on-farm risk factors for tail biting in pigs. *Appl. Anim. Behav. Sci.* 81:333-55. doi:10.1016/S0168-1591(02)00276-9.

Moyle, J. J., A. M. Fox, M. Arthur, M. Bynevelt and J. R. Burnett. 2007. Meta-analysis of neuropsychological symptoms of adolescents and adults with PKU. *Neuropsychol. Rev.* 17:91-101.

Mullan, S. 2009. An evaluation of including some welfare outcome measures within UK pig farm assurance schemes [Ph.D.]. University of Bristol. Available from: http://ethos.bl.uk/OrderDetails.do?uin=uk.bl.ethos.503854.

Mullan, S., W. J. Browne, S. A. Edwards, A. Butterworth, H. R. Whay and D. C. J. Main. 2009. The effect of sampling strategy on the estimated prevalence of welfare outcome measures on finishing pig farms. *Appl. Anim. Behav. Sci.* 119:39-48.

Mullan, S., S. A. Edwards, A. Butterworth, H. R. Whay and D. C. J. Main. 2011. Inter-observer reliability testing of pig welfare outcome measures proposed for inclusion within farm assurance schemes. *Vet. J.* 190:e100-9.

Munsterhjelm, C., O. A. T. Peltoniemi, M. Heinonen, O. Hälli, M. Karhapää and A. Valros. 2009. Experience of moderate bedding affects behaviour of growing pigs. *Appl. Anim. Behav. Sci.* 118:42-53.

Murray, G., K. Ashley and R. Kolesar. 2014. Drivers for animal welfare policies in Asia, the Far East and Oceania. *Rev. Sci. Tech. Int. Off. Epizoot.* 33:77-83.

NADIS. 2002. Pig Health - Tail Biting. Available from: http://www.nadis.org.uk/bulletins/tail-biting.aspx.

NADIS. 2003. Pig Health - Sunburn and Heatstroke/Heatstress. Available from: http://www.nadis.org.uk/bulletins/sunburn-and-heatstrokeheatstress.aspx?altTemplate=PDF.

Nasirahmadi, A., U. Richter, O. Hensel, S. Edwards and B. Sturm. 2015. Using machine vision for investigation of changes in pig group lying patterns. *Comput. Electron. Agric.* 119:184-90.

Nasirahmadi, A., O. Hensel, S. A. Edwards and B. Sturm. 2016. Automatic detection of mounting behaviours among pigs using image analysis. *Comput. Electron. Agric.* 124:295-302.

Newberry, R. C., D. G. M. Wood-Gush and J. W. Hall. 1988. Playful behaviour of piglets. *Behav. Processes.* 17:205-16.

Nielsen, E. K. and K. L. Ingvartsen. 2000. Effect of cereal type, disintegration method and pelleting on stomach content, weight and ulcers and performance in growing pigs. *Livest. Prod. Sci.* 66:271-82.

O'Connell, N. E., V. E. Beattie and R. N. Weatherup. 2004. Influence of group size during the post-weaning period on the performance and behaviour of pigs. *Livest. Prod. Sci.* 86:225-32.

Oczak, M., S. Viazzi, G. Ismayilova, L. T. Sonoda, N. Roulston, M. Fels, C. Bahr, J. Hartung, M. Guarino, D. Berckmans and E. Vranken. 2014. Classification of aggressive behaviour in pigs by activity index and multilayer feed forward neural network. *Biosyst. Eng.* 119:89-97.

OIE. 2016a. *Terrestrial Animal Health Code*. World Organization for Animal Health (OIE), Paris. Available from: http://www.oie.int/en/international-standard-setting/terrestrial-code/access-online/

OIE. 2016b. Report of the OIE ad hoc group on animal welfare and pig production systems. Available from: http://www.oie.int/fileadmin/SST/adhocreports/Animal%20welfare%20and%20pig%20production%20systems/AN/A_AW_Pig_Production_Systems_March_2016.pdf.

Omotosho, O. O., B. O. Emikpe, O. T. Lasisi, and T. A. Jarikre. 2016. Prevalence, distribution and pattern of gastric lesions in slaughtered pigs in south-western Nigeria. *Onderstepoort J. Vet. Res.* 83:6.

Pedersen, L. J., L. Holm, M. B. Jensen and E. Jørgensen. 2005. The strength of pigs' preferences for different rooting materials measured using concurrent schedules of reinforcement. *Appl. Anim. Behav. Sci.* 94:31-48.

Peet, B. nd. Using pig behaviour to optimize pen design. *Pig Site*. Available from: http://www.thepigsite.com/articles/928/using-pig-behaviour-to-optimize-pen-design/ (Accessed at 1 August 2017)

Popova, N. K. 2006. From genes to aggressive behavior: The role of serotonergic system. *BioEssays*. 28:495-503.

Potkins, Z. V., T. L. Lawrence and J. R. Thomlinson. 1989. Oesophagogastric parakeratosis in the growing pig: Effects of the physical form of barley-based diets and added fibre. *Res. Vet. Sci.* 47:60-7.

Puppe, B. 1998. Effects of familiarity and relatedness on agonistic pair relationships in newly mixed domestic pigs. *Appl. Anim. Behav. Sci.* 58:233-9.

Ramis, G., S. Gómez, F. Pallarés and A. Muñoz. 2005. Comparison of the severity of esophagogastric, lung and limb lesions at slaughter in pigs reared under standard and enriched conditions. *Anim. Welf.* 14:27-34.

Rasmussen, D. K., R. Weber and B. Wechsler. 2006. Effects of animal/feeding-place ratio on the behaviour and performance of fattening pigs fed via sensor-controlled liquid feeding. *Appl. Anim. Behav. Sci.* 98:45-53.

Rauw, W. M., E. J. Mayorga, S. Lei, J. C. M. Dekkers, J. F. Patience, N. K. Gabler, S. M. Lonergan and L. H. Baumgard. 2017. Effects of genetics on thermal regulatory responses to repeated heat stress exposure in pigs. *J. Anim. Sci.* 95:4-4.

Reimert, I., J. E. Bolhuis, B. Kemp and T. B. Rodenburg. 2013. Indicators of positive and negative emotions and emotional contagion in pigs. *Physiol. Behav.* 109:42-50.

Robertson, I. D., J. M. Accioly, K. M. Moore, S. J. Driesen, D. W. Pethick and D. J. Hampson. 2002. Risk factors for gastric ulcers in Australian pigs at slaughter. *Prev. Vet. Med.* 53:293-303.

Ruiterkamp, W. A. 1985. Het gedrag van mestvarkens in relatie tot huisvesting. Available from: https://bib.vetmed.fu-berlin.de/ResourceList/details/201962

Rydhmer, L., G. Zamaratskaia, H. K. Andersson, B. Algers, R. Guillemet and K. Lundström. 2006. Aggressive and sexual behaviour of growing and finishing pigs reared in groups, without castration. *Acta Agric. Scand. Sect. – Anim. Sci.* 56:109-19.

Rydhmer, L., K. Lundström and K. Andersson. 2010. Immunocastration reduces aggressive and sexual behaviour in male pigs. *Animal* 4:965-72.

Satterthwaite, D., G. McGranahan and C. Tacoli. 2010. Urbanization and its implications for food and farming. *Philos. Trans. R. Soc. B Biol. Sci.* 365:2809-20.

Schaefer, A. L., M. O. Salomons, A. K. W. Tong, A. P. Sather, and P. Lepage. 1990. The effect of environment enrichment on aggression in newly weaned pigs. *Appl. Anim. Behav. Sci.* 27:41-52.

Schmolke, S. A., Y. Z. Li and H. W. Gonyou. 2004. Effects of group size on social behavior following regrouping of growing-finishing pigs. *Appl. Anim. Behav. Sci.* 88:27-38.

Schön, P., B. Puppe and G. Manteuffel. 2004. Automated recording of stress vocalisations as a tool to document impaired welfare in pigs. *Anim. Welf.* 13:105-10.

Schrøder-Petersen, D. L., and H. B. Simonsen. 2001. Tail Biting in Pigs. *Vet. J.* 162:196-210.

Scott, K., D. J. Chennells, F. M. Campbell, B. Hunt, D. Armstrong, L. Taylor, B. P. Gill and S. A. Edwards. 2006. The welfare of finishing pigs in two contrasting housing systems: Fully-slatted versus straw-bedded accommodation. *Livest. Sci.* 103:104-15.

Spoolder, H. a. M., S. A. Edwards and S. Corning. 1999. Effects of group size and feeder space allowance on welfare in finishing pigs. *Anim. Sci.* 69:481-9.

Statham, P., L. Green and M. Mendl. 2011. A longitudinal study of the effects of providing straw at different stages of life on tail-biting and other behaviour in commercially housed pigs. *Appl. Anim. Behav. Sci.* 134:100-8.

Stavrakakis, S. 2014. *Biomechanical Studies of Locomotion in Pigs*. Newcastle University, Newcastle. Available from: https://theses.ncl.ac.uk/dspace/handle/10443/2510.

Stavrakakis, S., J. H. Guy, O. M. E. Warlow, G. R. Johnson and S. A. Edwards. 2014. Walking kinematics of growing pigs associated with differences in musculoskeletal conformation, subjective gait score and osteochondrosis. *Livest. Sci.* 165:104-13.

Stavrakakis, S., W. Li, J. H. Guy, G. Morgan, G. Ushaw, G. R. Johnson and S. A. Edwards. 2015. Validity of the Microsoft Kinect sensor for assessment of normal walking patterns in pigs. *Comput. Electron. Agric.* 117:1-7.

Stolba, A. and D. G. M. Wood-Gush. 1989. The behaviour of pigs in a semi-natural environment. *Anim. Sci.* 48:419-25.

Stookey, J. M. and H. W. Gonyou. 1994. The effects of regrouping on behavioral and production parameters in finishing swine. *J. Anim. Sci.* 72:2804-11.

Studnitz, M., M. B. Jensen, and L. J. Pedersen. 2007. Why do pigs root and in what will they root?: A review on the exploratory behaviour of pigs in relation to environmental enrichment. *Appl. Anim. Behav. Sci.* 107:183-97.

Stukenborg, A., I. Traulsen, B. Puppe, U. Presuhn and J. Krieter. 2011. Agonistic behaviour after mixing in pigs under commercial farm conditions. *Appl. Anim. Behav. Sci.* 129:28-35.

Swaby, H. and N. G. Gregory. 2012. A note on the frequency of gastric ulcers detected during post-mortem examination at a pig abattoir. *Meat Sci.* 90:269-71. doi:10.1016/j.meatsci.2011.06.015.

Taylor, N. R., D. C. J. Main, M. Mendl and S. A. Edwards. 2010. Tail-biting: A new perspective. *Vet. J.* 186:137-47.

Taylor, N. R., R. M. A. Parker, M. Mendl, S. A. Edwards and D. C. J. Main. 2012. Prevalence of risk factors for tail biting on commercial farms and intervention strategies. *Vet. J.* 194:77-83.

Telkänranta, H., J. N. Marchant-Forde and A. E. Valros. 2016. Tear staining in pigs: A potential tool for welfare assessment on commercial farms. *Anim. Int. J. Anim. Biosci.* 10:318-25.

The Pig Site. nd. Salt poisoning, water deprivation. Pig Site. Available from: http://www.thepigsite.com/pighealth/article/525/salt-poisoning-water-deprivation/ (Accessed at 1 August 2017).

Turner, S. P., D. J. Allcroft and S. A. Edwards. 2003. Housing pigs in large social groups: A review of implications for performance and other economic traits. *Livest. Prod. Sci.* 82:39-51.

Turner, S. P., S. A. Edwards and V. C. Bland. 1999. The influence of drinker allocation and group size on the drinking behaviour, welfare and production of growing pigs. *Anim. Sci.* 68:617-24.

Turner, S. P., M. J. Farnworth, I. M. S. White, S. Brotherstone, M. Mendl, P. Knap, P. Penny and A. B. Lawrence. 2006. The accumulation of skin lesions and their use as a predictor of individual aggressiveness in pigs. *Appl. Anim. Behav. Sci.* 96:245-59.

Turner, S. P., G. W. Horgan and S. A. Edwards. 2001. Effect of social group size on aggressive behaviour between unacquainted domestic pigs. *Appl. Anim. Behav. Sci.* 74:203-15.

Van de Weerd, H. A., C. M. Docking, J. E. L. Day, P. J. Avery and S. A. Edwards. 2003. A systematic approach towards developing environmental enrichment for pigs. *Appl. Anim. Behav. Sci.* 84:101-18.

van de Weerd, H. A. and J. E. L. Day. 2009. A review of environmental enrichment for pigs housed in intensive housing systems. *Appl. Anim. Behav. Sci.* 116:1-20.

van den Berg, A., F. Brülisauer and G. Regula. 2005. Prävalenz von Veränderungen der kutanen Magenschleimhaut bei Schlachtschweinen in der Schweiz. *Schweiz. Arch. Für Tierheilkd.* 147:297-303.

Vandermeulen, J., C. Bahr, E. Tullo, I. Fontana, S. Ott, M. Kashiha, M. Guarino, C. P. H. Moons, F. a. M. Tuyttens, T. A. Niewold and D. Berckmans. 2015. Discerning pig screams in production environments. *PLoS ONE.* 10:e0123111.

Vermeer, H. M., K. H. de Greef and H. W. J. Houwers. 2014. Space allowance and pen size affect welfare indicators and performance of growing pigs under Comfort Class conditions. *Livest. Sci.* 159:79-86.

Webb, N. G. 1984. Compressive stress on, and the strength of, the inner and outer digits of pigs' feet, and the implications for injury and floor design. *J. Agric. Eng. Res.* 30:71-80. Available from: http://agris.fao.org/agris-search/search. do?recordID=US201302564478.

Webb, N. G. and C. Nilsson. 1983. Flooring and injury – an overview. In: Baxter, S. H., Baxter, M. R. and McCormick, J. A.C. (Eds), *Farm Animal Housing and Welfare*. Nijhoff, The Hague, pp. 226-59.

Whay, H., C. Leeb, D. Main, L. Green and A. Webster. 2007. Preliminary assessment of finishing pig welfare using animal-based measurements. *Anim. Welf.* 16:209-11.

Wondra, K. J., J. D. Hancock, K. C. Behnke and R. H. Hines. 1995a. Effects of dietary buffers on growth performance, nutrient digestibility, and stomach morphology in finishing pigs. *J. Anim. Sci.* 73:414-20.

Wondra, K. J., J. D. Hancock, K. C. Behnke, R. H. Hines and C. R. Stark. 1995b. Effects of particle size and pelleting on growth performance, nutrient digestibility, and stomach morphology in finishing pigs. *J. Anim. Sci.* 73:757-63.

Woods, A. 2012. Rethinking the History of Modern Agriculture: British Pig Production, c.1910-65. *Twent. Century Br. Hist.* 23:165-91.

Wright, A. J. 2012. *Animal Welfare Assessment in Veterinary Education: Its Theory and Practical Application to Domestic Pigs (Sus scrofa domestica)*. University of London, London.

Yeates, J. W. and D. C. J. Main. 2008. Assessment of positive welfare: A review. *Vet. J.* 175:293-300.

Zijlstra, R. T. and E. Beltranena. 2013. Swine convert co-products from food and biofuel industries into animal protein for food. *Anim. Front.* 3:48-53.

Chapter 3

Optimising pig welfare during transport, lairage and slaughter

Luigi Faucitano, Agriculture and Agri-Food Canada, Canada; and Antonio Velarde, Institute of Agrifood Research and Technology, Spain

1 Introduction

Stress imposed on pigs during transport, in lairage and at slaughter is both an animal welfare and a meat quality issue. Studies revealed that poor handling practices and transport conditions and wrong facility design at the slaughter plant both, individually and/or additively, can contribute to the loss of profits due to animal losses (dead-on-arrival or DOA and downers), reduction in carcass value due to weight losses and skin bruises and meat quality defects due to abnormal *post-mortem* muscle acidification (Schwartzkopf-Genswein et al., 2012; Faucitano, 2018; Rioja-Lang et al., 2019).

The responsibility for animal losses during transport may either be equally shared among the producer, who must guarantee proper preparation of animals (i.e. feed withdrawal) prior to transport and handling of pigs at loading, the trucker and the abattoir (in the case of integrated production systems) or be bore alone by the trucker, whereas the abattoir is responsible alone for the optimisation of lairage and slaughter conditions (layout, ambient control and handling systems) in order to let pigs recover from the stress of handling and transport and ensure optimal and uniform carcass and meat quality.

Training of handling crews and the application of rewards and fines have become of paramount importance to improve handling and reduce animal

http://dx.doi.org/10.19103/AS.2020.0081.07

losses (Correa, 2011; Rocha et al., 2016; Dalla Costa et al., 2019). A Canadian integrated company reported a decrease from 0.3% to 0.01% in the incidence of fatigued pigs on arrival at the abattoir by applying a new programme of payment with incentives to handlers and truckers as a reward for their willingness to slow down the loading/unloading speed rate to 100 pigs/h (Correa, 2011). In contrast, fines up to $6000 under Canadian Food Inspection Agency regulations are applied to Canadian truckers for having –three to four dead pigs in the truck load (Faucitano, 2018). Grandin (2018) recently reported that the decision of an abattoir management in the US to apply a $25 handling fee for each pig arriving in a non-ambulatory condition resulted in a great reduction in the occurrence of downers.

This chapter will focus on the handling practices to be applied during the transportation of pigs (market weight and culled sows) to slaughter, as this phase is considered as the most stressful of the whole preslaughter period, and in lairage and at stunning/slaughter at the abattoir, aiming at limiting their impact on animal losses and pork quality.

2 Welfare during transport

The situation when a pig is in transit is considered a major stressor and may have deleterious effects on its health, its well-being, carcass yield and ultimately pork quality. Major sources of stress during transport are, among others, vehicle design, the time spent in the truck and the space allowed to lie down.

2.1 Vehicle design

Vehicle design features that may influence the welfare of pigs during transport include the loading/unloading system (ramps or hydraulic decks), the compartment location and the microclimate control (Faucitano and Goumon, 2018; Rioja-Lang et al., 2019).

The presence of fixed decks and ramps within the vehicle makes the procedures of loading and unloading more difficult, resulting in a higher risk of animal losses (Barton-Gade et al., 2007; Correa et al., 2013) and poor pork quality, that is, either pale, soft and exudative (PSE) or pale or dark, firm and dry (DFD) pork (Guàrdia et al., 2004; Correa et al., 2013, 2014). The North American pot-belly (PB) trailer is an example of the vehicle design featuring two to five, often steep (up to 32° slope), ramps (see Fig. 1) and 180° turns that pigs have to negotiate during the loading and unloading process. These trailer features have been associated with a lower ease of handling as showed by the greater use of electric prods and longer handling time (Ritter et al., 2008; Torrey et al., 2013a,b; Weschenfelder et al., 2013b), eventually resulting in a greater proportion of DOA and fatigued pigs on arrival at the abattoir compared with other trailer models featuring hydraulic decks (Ritter et al., 2008; Sutherland

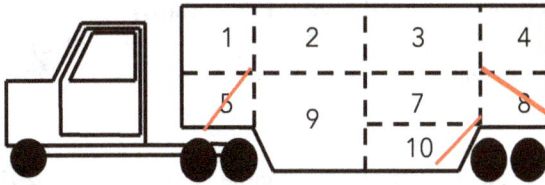

Figure 1 Position of compartments and internal ramps (solid red lines) inside a pot-belly trailer (modified from Correa et al., 2014). The represented model is a dual-purpose pot-belly trailer (for cattle and pig transport).

et al., 2009; Kephart et al., 2010; Weschenfelder et al., 2012; Correa et al., 2013).

The presence of ramps is one of the major contributors to the handling problems reported during loading and unloading in PB trailers (see Figs. 1 and 2; Faucitano and Goumon, 2018; Rioja-Lang et al., 2019). It is well known, in fact, that negotiating a ramp slope from 0° to 45° represents a significant physical and psychological challenging experience for pigs, as showed by the increased heart rate, frequency of turn backs and baulking, skin bruises and

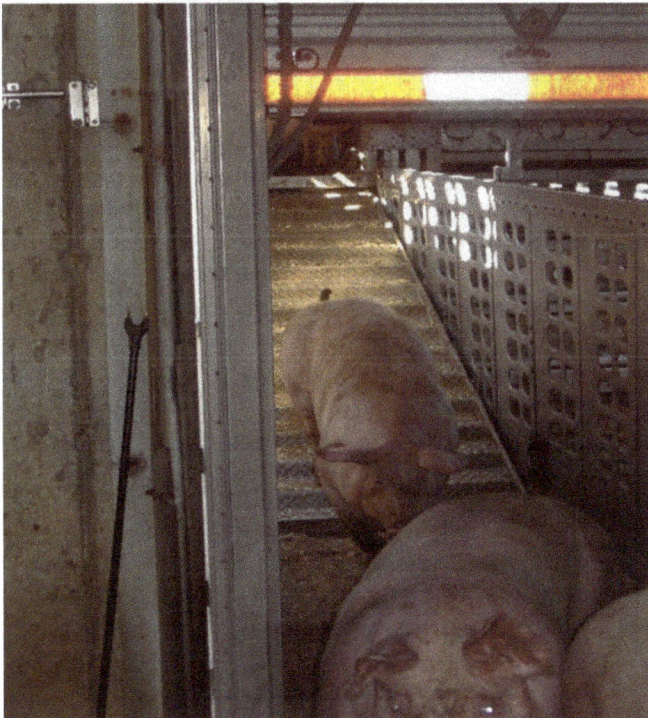

Figure 2 Unloading from a pot-belly trailer through an internal ramp (L. Faucitano, AAFC, Canada).

handling time (Van Putten and Elshof, 1978; Warriss et al., 1991; Dalla Costa et al., 2007; Goumon et al., 2013b; Garcia and McGlone, 2015). Recently, Dalla Costa et al. (2019) compared ramp slopes of < 20° versus > 20° and reported a four-fold greater risk for DOA when steeper ramps were used. Furthermore, steep ramps (up to 26°) are also challenging for the handler, as showed by the increased heart rate and difficulty to handle (Goumon et al., 2013b). Current guidelines do not recommend to use ramps steeper than 20° for fixed ramps or than 25° for adjustable ramps (SCAHAW, 2011; TQA, 2016). However, as market pigs have become heavier (from 113 kg to 130 kg from 2000 to 2010 in Canada; Correa, 2011) and difficult to move (Bertol et al., 2011; Rocha et al., 2016), recent works recommend the maximum ramp slope to be reduced to 15° (Grandin, 2012; Goumon and Faucitano, 2017).

2.2 Animal position in the truck

The impact of the deck and/or compartment position on the vehicle pigs' welfare during transport and meat quality can be explained by the differences in the ease of handling (due to the presence of ramps), airflow rate, vibration rate and loading order (Faucitano and Goumon, 2018). European and Canadian studies showed that, when compared with the middle deck, pigs transported in the top and/or bottom decks presented higher body temperature, exsanguination blood cortisol levels, dehydration rates, carcass lesion score and incidence of PSE and exudative or pale/dark pork (Lambooij et al., 1985; Lambooij and Engel, 1991; Barton-Gade et al., 1996; Correa et al., 2013). The front and rear top and bottom rear compartments can be also critical, as showed by the increased core body temperatures after loading and during transport and higher heart rate after loading and longer unloading time (Goumon et al., 2013a; Torrey et al., 2013b; Conte et al., 2015). A greater risk for DFD pork production has been reported in pigs transported in the middle front compartment of a PB trailer (Correa et al., 2014), especially after long transportation (18 vs. 6 and 12 h) in winter (Scheeren et al., 2014).

2.3 Control of the microclimate within the transport vehicle

Generally, greater animal losses are reported during summer hauls (Vecerek et al., 2006; Werner et al., 2007; Haley et al., 2008; Correa et al., 2013; Vitali et al., 2014), with greater risk being recorded at ambient temperatures either between 17°C and 20°C (Warriss and Brown, 1994; Sutherland et al., 2009; Haley et al., 2008, 2010; Kephart et al., 2010) or between 29°C and 33°C (Peterson et al., 2017).

Within a vehicle the internal temperature can increase by almost 1°C for each 1°C increase in the environmental temperature (Dewey et al., 2009). The internal temperature increases more, up to 3-4°C rise in 5 min at a rate of approximately 1°C/min, in passively ventilated vehicles held in a stationary situation (Lambooij, 1988; Xiong et al., 2015). However, when passively ventilated trailers are compared, the environment is warmer in the PB trailer compared with the triple-deck flat-deck (FD) trailer while stationary or moving (Ritter et al., 2008; Weschenfelder et al., 2012; Faucitano and Goumon, 2018). The variation in the internal thermal conditions between the two trailer models has been explained by the difference in the side openings type (punch for the PB vs. slatted for the FD trailer) influencing the air flow inside the vehicle (Weschenfelder et al., 2012). More specifically, within the PB trailer the highest temperature peak (30.3°C) has been recorded in the bottom deck during transport (Brown et al., 2011), while the greater gradient (up to +11°C) between internal trailer and external ambient temperatures has been recorded in the middle and bottom front compartments during stops (Weschenfelder et al., 2012, 2013; Fox et al., 2014).

To ensure thermal comfort and reduce animal losses during vehicle stops, pigs should be cooled off by fan-assisted ventilation, water sprinkling/misting or the two systems combined. Research showed that when applied for 5 min after loading and before leaving the farm and at the end of the wait before unloading at the abattoir at ambient temperatures of 20°C and greater, water sprinkling reduced pigs' fatigue (as assessed by the blood lactate concentration) at slaughter and pork exudation, especially in pigs transported in the middle front and rear compartments (Nannoni et al., 2014). However, water sprinkling combined with insufficient ventilation can also result in an increased difference in humidity levels (up to +7.5%) between the trailer interior and the external environment (Fox et al., 2014), preventing efficient evaporative cooling in pigs. Pereira et al. (2018) showed that the use of fan-misting bank installed in the abattoir yard (see Fig. 3) for 30 min under warm ambient conditions (up to 28.1°C) was efficient to improve the quality of the internal environment of a PB trailer (lower temperature–humidity index or THI) and the pigs' thermal comfort (lower body heat loss) while waiting before unloading.

A greater proportion of DOAs and non-ambulatory pigs on arrival at the plant has been also reported in winter than in summer (Guàrdia et al., 1996; Sutherland et al., 2009), with a higher risk of death at 4-10°C than at 12-26°C (Peterson et al., 2017). Likely causes of the greater animal losses in winter compared with other seasons may be the more difficult animal handling through the icy internal ramps at loading and unloading (Torrey et al., 2013a,b) and insufficient bedding of the trailer floor, resulting in more pigs standing during transport to avoid the contact with the cold aluminium floor surface (Goumon et al., 2013a). The presence of slippery ramps can result in more slips and falls at loading and unloading (Torrey et al., 2013b), greater heart rates during transport and unloading (Goumon et al., 2013a) and increased blood creatine kinase (CK) and lactate concentrations at slaughter (Correa et al., 2014).

Figure 3 Operation of a fan-misting bank on pigs kept in a pot-belly trailer during their wait before unloading (L. Faucitano, AAFC, Canada).

In winter, the thermal comfort of pigs in the truck can be controlled by partially closing the ventilation openings (boarding) in order to reduce air flow and by adding 5-cm layer of Styrofoam to the vehicle top ceiling (Gonyou and Brown, 2012) and bedding on the truck floor (TQA, 2016). Within the air temperature range between 15°C and ≤ −12°C, trucks should be utilising from 25% to 95% boarding to prevent DOAs (McGlone et al., 2014a; TQA, 2016).

In winter, no more than six bales of bedding should be sufficient to avoid pigs to be in contact with the cold floor surface of the trailer during transport and reduce the rate of DOA or downers and the number of carcass lesions, including frostbites (Goumon et al., 2013a; McGlone et al., 2014b; Scheeren et al., 2014). The number of bales should be reduced to three in summer (McGlone et al., 2014b).

2.4 Journey time

Both long and short journeys can affect the welfare of slaughter pigs and mortality rate during transport (Rioja-Lang et al., 2019). Some studies reported an increased risk of animal and body weight losses (Warriss et al., 1990; Sutherland et al., 2009), fatigue and dehydration, as shown by the higher blood glucose, lactate and haematocrit levels at slaughter (Brown et al., 1999; Mota-Rojas et al., 2006; Becerril-Herrera et al., 2010; Sommavilla et al., 2017), and DFD pork production (Leheska et al., 2003; Mota-Rojas et al., 2006) with transport time longer than 4 h and up to 24 h. In contrast, in other studies shorter journeys (≤2 h) resulted in a lower ease of handling at the abattoir, greater concentrations of cortisol and lactate in exsanguination blood and a higher risk of PSE pork production (Faucitano and Lambooij, 2019). A possible reason for the detrimental effect of short journeys is that the pigs could not recover from the stress of handling experienced at loading upon arrival at the abattoir (Weschenfelder et al., 2013).

However, the effects of travel time on pig welfare can be additive to that of other concurrent factors, such as the time on fasting, season of the year, type of vehicle, loading density and pig genetic background (Rioja-Lang et al., 2019). Fasting pigs on farm for 12–18 h before loading reduces the risk of mortality during transport, regardless of the journey duration (up to 24 h), while in unfasted pigs this risk increases for journeys up to 8 h (Averós et al., 2008). In summer increased risk for DOA has been reported in hauls longer than 2 h (Vitali et al., 2014), while greater core body temperature, heart rate and incidence of DFD pork production have been recorded in pigs transported for 18 h in winter than in summer (Goumon et al., 2013a; Scheeren et al., 2014). The latter results may be caused by the longer exposure of pigs to cold stress. In transportation trials of different durations (45 min and 7 h), using both PB and FD trailers, Weschenfelder et al. (2013) reported an increased level of fatigue (based on exsanguination blood lactate concentration) at the time of slaughter in stress-susceptible pigs (Hal[Nn]) hauled for a shorter time with the PB trailer compared with a FD trailer equipped with semi-hydraulic middle and top decks.

2.5 Space allowance

The recommended loading density for pigs during transport involves a trade-off between the economic pressure to increase the loading density in order to maximise profit from a single journey and the welfare of animals during transport (Rioja-Lang et al., 2019). However, particular attention must be paid to this transport variable considering its significant contribution to animal losses. In a US survey of more than 12 000 loads, Fitzgerald et al. (2009) reported that trailer density accounted for the largest portion of the variation in the total animal losses, that is, dead, fatigued and injured pigs, compared to other variables, such as the driver, handling crew, THI, wind speed, loading duration and wait time before unloading.

The EU legislation (Council Regulation (EC) No 1099/2009, 2009) is based on the evidence that when the loading density is higher than 235 kg/m^2 or lower than 0.425 m^2/100 kg not all pigs are able to lie down to rest and cannot rest as they are pushed to continually change their position (Lambooij et al., 1985; Lambooij, 2014). This uncomfortable situation has been associated with increased mortality rates and a higher number of non-ambulatory pigs on arrival at the plant (Ritter et al., 2006). Fitzgerald et al. (2009) reported that increasing the density from 212.4 kg/m^2 to 338.6 kg/m^2 corresponded to a 7.5-fold increase in animal losses on arrival at the slaughter plant. However, both low (0.50 m^2/100 kg) and high density (0.33 m^2/100 kg) during transport can cause physical stress, resulting in muscle fatigue and glycogen depletion, and a greater risk of DFD pork production (Lambooij et al., 1985; Guàrdia et al., 2005). In a low-density situation, physical stress is caused by the attempt of pigs to maintain their balance and cope with the unexpected and sudden movements of the truck or by fighting due to their greater freedom to move around in the compartment (Barton-Gade and Christensen, 1998; Guàrdia et al., 2005). In contrast, providing less space may fatigue pigs due to the frequent disturbance of lying animals by those seeking a place to rest and the difficulty of standing pigs to maintain their balance while the vehicle is moving (Lambooij and Engel, 1991).

The application of loading densities should be adjusted according to the body weight, ambient conditions and travel time. A recent Italian study reported a greater risk of DOAs when the EU-recommended density (235 kg/m²) is applied for the transport of 160 kg pigs, which would correspond to a loading density of 0.7 pigs/m² (Nannoni et al., 2016). Because of their different physical needs and greater susceptibility to heat stress, due to the decreased heat dissipation rate (Renaudeau et al., 2011), the minimum recommended truck space required by heavier-market-weight pigs (from 114 kg to 182 kg) should be increased from 0.40 m²/pig to 0.61 m²/pig in winter and from 0.46 m²/pig to 0.65 m²/pig in summer (Grandin, 2017b). If the load size is simply determined by the number of animals that need to be shipped, then heavier pigs may be too packed on the truck (Zurbrigg et al., 2017). During hauls of 140-kg weight pigs, the reduction of the number of pigs from 3/m² to 2/m² has been associated with a decreased risk of animal losses from 1% to 0.2% (Fitzgerald et al., 2009).

During hot and humid ambient conditions, it is recommended to provide 15% to 25% more space (CARC, 2001; https://ec.europa.eu/food/animals/w elfare/practice/transport_en). Indeed, transport losses are lower when more truck floor space (0.46 m²/100 kg vs. 0.39 m²/100 kg) is provided in summer (Ritter et al., 2012). However, greater space allowance (0.50 m²/100 kg vs. 0.25 m²/100 kg) may increase the risk for skin lesions (+28.2%) in this season (Guàrdia et al., 2009). In winter, the risk of skin lesions is higher (9.7%) at high loading density (0.25 m²/100 kg) due to the effect of huddling to keep the body temperature and fighting/mounting to seek for space to lie down (Guàrdia et al., 2009).

Pilcher et al. (2011) showed that the increase of floor space (from 0.40-0.49 m²/100 kg to 0.52 m²/100 kg) helps reduce the incidence of fatigued pigs on arrival at the plant after short transport (<1 h) compared with longer journeys (3 h) as pigs can get adapted more to a long journey (550 km) as showed by the higher proportion of pigs lying on the truck floor and reduced body temperature and heart rate (Gerritzen et al., 2013). However, short journeys (1 h vs. 3 h) at higher loading densities (0.25 m²/100 kg vs. 0.50 m²/100 kg) have been also associated with a decreased risk of PSE pork production (Guàrdia et al., 2004). Based on this evidence, in order to prevent this meat quality defect, the EU-recommended space allowance of 0.425 m²/100 kg should be only applied for journeys longer than 3 h. These results may be explained by the fact that at the departure from the farm pigs do not lie down immediately but still stand exploring and getting adapted to the novel environment. They remain so for the first few kilometres (0.5 h) to better cope with the fear and stress (as showed by the increased heart rate up to 220 beats/min) caused by the frequent truck accelerations/decelerations and vibrations generated by driving on the initial country unpaved roads (Chevillon, 2001). It can take up to 2 h of transport for the pigs to settle down in the truck (Lambooij et al., 1985; Barton-Gade and Christensen, 1998). In these conditions, providing pigs with less space would help them keep their balance by holding each other while the vehicle negotiates bends or poor road surfaces (Barton-Gade and Christensen, 1998).

2.6 Transport of culled sows

In pig production, sows are most likely culled due to poor body condition, lameness or failure to rebreed (Baloyh et al., 2015; Zhao et al., 2015; Grandin, 2016a). Such defects, along with greater mortality, are mostly reported at the largest farms (Koketsu, 2000).

Because of poor body and ambulatory conditions, cull sows may hardly walk and be loaded onto the truck, resulting in a greater risk of arriving fatigued, lame and even with worst body conditions, or dead, at the destination (i.e. buying station or slaughter plant) compared with market-weight pigs (Malena et al., 2007; McGee et al., 2016; Peterson et al., 2017). In a recent observational study on 47 loads of sows transported from the farm to slaughter, Thodberg et al. (2019) observed a deterioration of clinical signs, such as vulva and udder ulcers, skin lesions, wounds and torn hoofs, on arrival at the slaughter plant. These observations support the recommendation for producers to select fit sows and boars for transportation (Bench et al., 2008; Grandin et al., 2016a).

The effects of transportation on sow welfare can be exacerbated by the very long distance cull sows have to travel from the farm or assembly yard to the abattoir and ambient conditions. A recent study surveying the cull sow market network in the US reported that sows could be marketed between 24 and 120 h and travel straight to slaughter for an average of 1057 km prior to harvest (Blair and Lowe, 2019). Peterson et al. (2017) reported that for cull sows the risk to die during transport may be 1.93 and 0.81 times higher at outdoor ambient temperatures ranging from 29°C to 33°C and from 4°C to 10°C, respectively, compared to 12°C to 26°C.

Transport duration (4 h on average, ranging from <1 to 8 h), temperature in the truck (14°C on average, ranging from 3°C to 26°C) and duration of stops and wait before unloading at the slaughter plant (0.5 h on average for both, ranging from 0 h to 3 h and from 0 min to 75 min, respectively) have also been

Published by Burleigh Dodds Science Publishing Limited, 2021.

identified as the most important causing factors for the deterioration of clinical signs in cull sows on arrival at the abattoir (Thodberg et al., 2019).

3 Welfare in lairage

The purpose of lairage is to give stressed or fatigued animals an opportunity to recover from the stress of transport and previous handling in order to produce better meat quality (Faucitano, 2010, 2018; Gallo et al., 2016).

The recovery rate of pigs in lairage and the related economic losses due to poor carcass and meat quality depend on lairage time, the quality of the handling systems and ambient control (Faucitano, 2010; Gallo et al., 2016).

3.1 Lairage time

Two to three hours is usually the recommended rest time to allow the pigs to recover their physiological condition after transport and handling and to ensure the production of good-quality pork (Warriss, 2003). Unless required by the adverse ambient conditions, such as too high ambient temperature (>30°C; Fraqueza et al., 1998) or ammonia levels (>10 ppm; Weeks, 2008), short lairage intervals (<1 h) should not be applied as they may result in a higher muscle temperature and lactate level at slaughter, resulting in an increased incidence of PSE pork production (Fraqueza et al., 1998; Shen et al., 2006). However, while longer lairage time helps reduce the risk of PSE pork production (Guàrdia et al., 2005), extending lairage time from 3 h to overnight increases the risk of DFD pork production (Guàrdia et al., 2005) due to the depletion of the muscle glycogen content at slaughter caused by the combined effect of fasting and fighting within mixed groups of unfamiliar pigs (Nanni Costa et al., 2002; Guàrdia et al., 2009; Dalla Costa et al., 2016). Fighting-related skin lesion scores, in fact, increase with the lairage time (Warriss, 1996; Faucitano, 2001, 2010; Bottaccini et al., 2018), with almost a two-fold higher risk in pigs kept in lairage for 15 h compared to 3 h (Guàrdia et al., 2009).

3.2 Mixing

Under commercial conditions, mixing groups of unfamiliar pigs in lairage is a common practice because of the difference in size between the truck compartment and lairage pen, despite the risk of fighting, in terms of biting, pushing and head knocks (Rabaste et al., 2007) between pigs, and carcass value downgrading due to severe skin lesions (Faucitano, 2001). However, some strategies can be applied to limit fighting in lairage, such as keeping pigs in smaller groups (10 pigs/group vs. 30 pigs/group; Rabaste et al., 2007) and reducing the space allowance (0.85 m²/pig to 0.42 m²/pig; Moss, 1978; Geverink et al., 1996; Weeks, 2008). Anecdotally, some European plants managed to reduce fighting by enriching the pen with some corn kernels scattered on the floor that may distract pigs or sprinkling the back of pigs with vinegar that apparently mask the individual smell of each pig (Eyes on Animals, 2019).

The intensity and duration of fighting in lairage can be also aggravated by pig gender, with boars being more aggressive than gilts and barrows (Warriss and Brown, 1985) and immunocastrates fighting two-fold more than surgical castrates, especially when fed ractopamine during the last 28 days of the finishing period (Rocha et al., 2013). In a more recent study, Wesoly et al. (2015) reported an inverse relationship between testicular function and skin lesion score within mixed groups of boars, with high-ranking boars presenting lower skin lesions compared with low-ranked ones that are mounted more frequently (Rydhmer et al., 2006).

3.3 Ambient control

The control of temperatures between 15°C and 18°C and relative humidities (RH) between 59% and 65% is recommended to ensure the thermal comfort of pigs in lairage (Honkavaara, 1989). When these environmental conditions are not respected, pigs can either suffer from cold stress (shivering and huddling), which may result in DFD meat due to muscle energy depletion to maintain a constant body temperature (Knowles et al., 1998) or heat stress, as showed by increased panting, especially when they are kept at hot (>30°C) and humid (RH > 80%) conditions (Santos et al., 1997). According to Grandin (2012), death losses in lairage, also called 'dead-in-pen', almost doubled at temperatures above 32°C compared with 16°C.

Water spraying on pigs in the lairage pen helps reduce the heat-related respiration rate (Huynh et al., 2006), the number of dead-in-pen (Vitali et al., 2014) and the incidence of PSE pork production through a 2°C drop in the muscle mass temperature (Long and Tarrant, 1990). However, at temperatures below 5°C, showering is not recommended as it causes animal shivering and may lead to DFD pork due to muscle energy depletion to maintain a constant body temperature (Knowles et al., 1998). Ivn order to remove the excessive humidity produced by the application of sprinkling/misting systems as well as to control the concentration of noxious gases, for example, ammonia (Weeks, 2008), this practice must be combined with a proper ventilation (135 m^2 h^{-1}; Brent, 1986; Weeks, 2008). Vitali et al. (2014) reported a lower risk of death in lairages characterised, among others, by efficient ventilation through large open windows on the roof and side walls.

When compared to other livestock species (i.e. cattle and sheep), pig lairages are the loudest ones (Weeks et al., 2009), with the average noise level ranging from 76 dB to 108 dB and the highest peaks (120 dB) being recorded in the *peri-mortem* area (Talling et al., 1996; Rabaste et al., 2007). Excessive lairage noise produces a fear response in pigs, as showed by the number of pigs huddling in the pen looking for protection or escaping from the source of sound (Geverink et al., 1998), the increased heart rate and greater blood lactate, CK and cortisol levels at slaughter (Faucitano, 2010), all resulting in an increased production of PSE pork (Warriss et al., 1994; Van de Perre et al., 2010). Keeping the sound level lower than 85 dB in the *peri-mortem* area appears to reduce the risk for PSE meat (Vermeulen et al., 2015).

Lairage noise is mostly caused by gates clanging, operating machinery, echoes and pig vocalisation (Weeks, 2008; Weeks et al., 2009), although pigs appear to be more stressed by industrial sounds than the sounds of conspecifics (Geverink et al., 1998). To reduce the ambient sound level, some European plants replaced metal gates and fencing with plastic ones and modified the ceiling by decreasing the height and installing sound-absorbing materials (Eyes on Animals, 2019).

The movement of pigs in the alleys of the slaughter plant may be influenced by the lighting of these areas. Pigs are less reluctant to move from a dark area to a brightly lighted area (Van Putten and Elshof, 1978; Grandin, 1990; Tanida et al., 1996), making the use of moving tools less necessary. Grandin (2010) observed that the insufficient light (less than 160-215 lux) at the entrance of the stunning area increased the use of electric prods by 34%. The alleys must be consistently lighted to avoid shadows and contrasts in colour on the floor. It has been recently observed that the use of green lighting reduces shadows on the floor and improves the ease of handling through the alleys (Eyes on Animals, 2019; see Fig. 4).

3.4 Driving pigs to slaughter

The combination of a higher slaughter speed, poorly designed handling systems and the progressive passage from a free-moving group situation to a single line of aligned and restrained individuals during the short period of time between the exit from the lairage pen and stunning may result in a greater proportion of slips, jamming, backing-up and vocalisation (Warriss et al., 1994; Edwards et al., 2010, 2011; Van de Perre et al., 2010; Vermeulen et al., 2015; Rocha et al., 2016), and increased use of electric prods (Rocha et al., 2016). These behavioural responses have been associated with increased heart rate (up to 240 beats/min; Chevillon, 2001; Correa et al., 2010), blood lactate and CK levels at slaughter (Hambrecht et al., 2005; Edwards et al., 2010; Rocha et al., 2015), skin lesions scores (Rabaste et al., 2007) and exudative pork (Van der Wal et al., 1999; Hambrecht et al., 2005; Rabaste et al., 2007; Dokmanović et al., 2014; Rocha et al., 2016).

In the *peri-mortem* area critical factors are the entrance into the stun chute and the 'stop-start' forward movement of pigs towards the stunner,

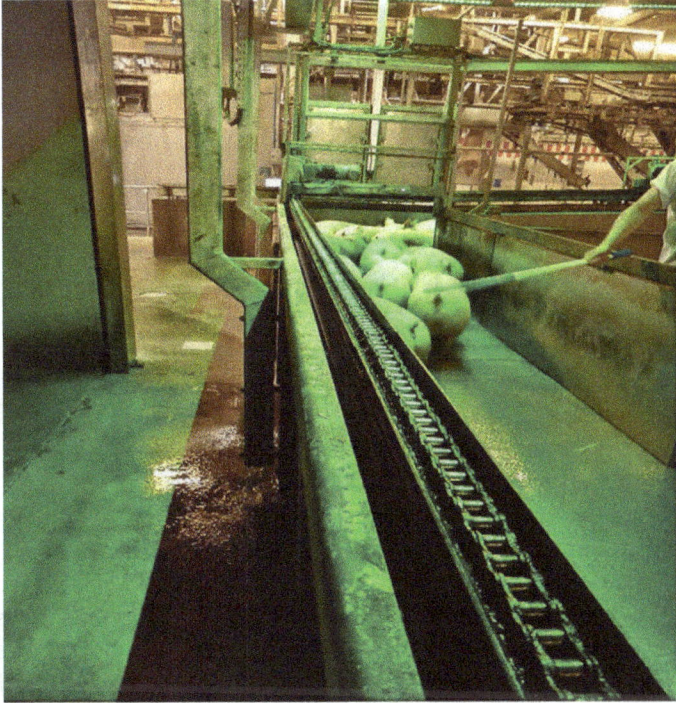

Figure 4 Green lighting reduces the presence of shadows on the alley floor preventing stops and backups slowing down the flow of pigs towards slaughter (Photo courtesy of Eyes on Animals, The Netherlands).

which are both observed in raceways feeding electrical and CO_2 stunners and which result in the frequent use of electric prods to encourage the pig movement (Faucitano, 2010). The batch flow system or curved raceway has been observed to work well, easing the flow of pigs into the single/double electric stun raceway, as showed by the lower number of reluctant-to-move-forward pigs and the use of electric prods compared with the single file chute entrance (Edwards et al., 2010). However, this system must be well designed and managed, in that it must not have an abrupt entrance and should not be filled with pigs for more than 75% of its capacity (ideally 50%) to prevent jamming and overlaps in the crowd pen (Grandin, 2017b; see Fig. 5). The entrance of pigs into a CO_2 gas stunner has been significantly improved by the group-wise stunning system (Christensen and Barton-Gade, 1997), where pigs are moved forwards in groups of 15 pigs using pushing gates and are loaded in sub-groups of 5 into the cradle. This system proved to reduce the frequency of PSE (−3%) and blood-splashed pork (−5%) and bruised carcasses (−4%) due to reduced electric prodding and physical stress (Christensen and Barton-Gade, 1997).

Published by Burleigh Dodds Science Publishing Limited, 2021.

4 Welfare during stunning and slaughter

The last stage of pig production is the slaughter for human consumption. The slaughter process includes the restraining of the animal, the stunning application and the exsanguination.

4.1 Restrainer types

The purpose of restraint is to facilitate the correct application of the stunning and bleeding. The restrainer type varies depending of the stunning method. In electrical stunning, restraint is needed to facilitate the correct placement of the electrodes between the eye and the ear and to facilitate the uninterrupted flow of current. However, incorrect restraint may lead to ineffective stunning or bleeding but can also cause pain and distress in its own right. The restrainer should match the size of the pigs. It should be narrow enough that the animals cannot move backwards and forwards or turn around. Insecure restraint may cause struggling or escape attempts (Grandin, 2016b). On the other hand, a too narrow device will provide excessive force to the animal and will be painful and may cause injuries. Prolonged restraint time may cause fear and exacerbate insecure or excessive restraint. Therefore, no pig should enter the restrainer until equipment and personnel are ready to slaughter that animal.

Several automatic electrical stunning methods are currently available. One device consists of a V-type restrainer where each pig makes contact with the electrodes and receives the stunning current (Fig. 6). Both sides must run at the

Figure 5 Filling the crowd pen with pigs for more than 75% of its capacity prevents easy handling into the stunning chute (L. Faucitano, AAFC, Canada).

Published by Burleigh Dodds Science Publishing Limited, 2021.

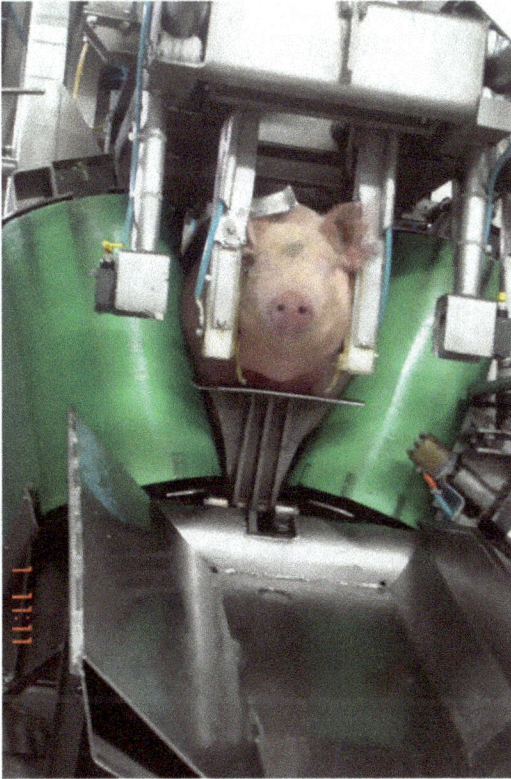

Figure 6 V-type restrainer (A. Velarde, IRTA, Spain).

same speed (Grandin, 2016b). When the two conveyors run at different speeds and their angle is too steep, the animals may struggle. A second method uses a conveyor belt system (Fig. 7). In this system, at the end of the restrainer, the nose of the pigs interrupts a beam of light, which activates the electrodes. In both systems, the animals are turned out and fall onto a table after the stunning.

Animal-based measures to assess pain, fear or distress of the pigs in the restrainer include slips or falls, struggling, escape attempts and vocalisation (Velarde and Dalmau, 2012; Grandin, 2016b).

Comparative studies showed that pigs restrained in the conveyor belt presented a lower heart rate (180 heartbeats/min vs. 220 heartbeats/min) and proportion of PSE pork production compared with those conveyed with the V-type restrainer (Griot et al., 2000).

On the other hand, the exposure to CO_2 at a high concentration does not require restraint as animals are stunned in groups (Velarde et al., 2000). In this case, gondolas for the gas stunning of pigs should not be overloaded, and animals should be able to stand without being on top of each other. Gondolas

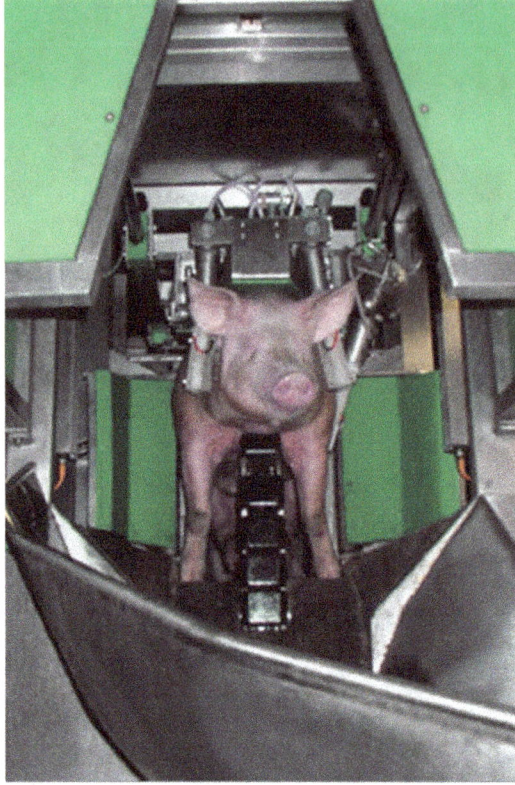

Figure 7 Conveyor belt system (A. Velarde, IRTA, Spain).

overloading is a criterion in animal welfare auditing protocols at slaughter plants (Grandin, 2017b) based on the observation of 25-50 gondolas, depending on the plant size. Audit scores are given on a gondola basis and range from excellent (no gondolas are overloaded) to serious problem (the person moving the animals forces one or more pigs to jump on top of the other pigs in the gondolas with an electric prod or by hitting, shoving or kicking).

4.2 Stunning methods

To spare any avoidable pain, distress or suffering, stunning before bleeding is a common practice around the world. Stunning is defined as any intentionally induced process which causes the loss of consciousness and sensibility without pain, including any process resulting in instantaneous death (Regulation No. 1099/2009). During the loss of consciousness, the animal is unable to perceive external stimuli and control its voluntary mobility and, therefore, does not respond to normal stimuli, including pain (EFSA, 2004). In order to fulfil the humane slaughter requirements, the duration of unconsciousness must be longer than the sum of time that lapses between the end of stun and the time to the onset of death due to bleeding. Therefore, unconsciousness should be monitored at different stages of the process: (1) during shackling and hoisting, (2) before sticking and (3) during bleeding (EFSA, 2013). The EFSA AHAW panel has developed toolboxes of welfare indicators for developing monitoring procedures at slaughterhouses for pigs (EFSA, 2013). The most commonly used methods for stunning pigs at slaughter are electrical stunning and exposure to CO_2 at high concentrations.

Electrical stunning (Fig. 7) involves the application of an electric current of sufficient magnitude to the brain such that a generalised epileptiform activity is induced similar to that recorded in humans during grand mal epileptic seizures (Croft, 1952; Hoenderken, 1978). This seizure-like state, immediately followed by an exhausted state, is suggestive of an immediate loss of consciousness and appears to be associated with a lack of sensory awareness, which lasts a finite period of time (Anil, 1991). An effective stun is characterised by the presence of all of tonic–clonic seizures, the loss of posture, apnoea and the absence of corneal reflex (Velarde and Dalmau, 2012). The main hazards preventing effective stunning are incorrect electrode placement, poor contact, dirty or corroded electrode, too low voltage/current or high frequency (EFSA, 2004).

Exposure to CO_2 at high concentrations is also used for the stunning of pigs. In this case, the loss of consciousness is not immediate (Rodríguez et al., 2008). Since CO_2 does not induce immediate loss of consciousness, inhalation of concentrations greater than 30% of CO_2 by volume in the atmospheric air causes aversion, irritation of the mucous membranes (that can be painful) and respiratory distress during the induction phase (Velarde et al., 2007). During CO_2 exposure, pigs show signs of aversion such as retreat attempts, headshaking, sneezing, breathlessness, freezing, escape attempts, gasping (a very deep breath through a gasping-open mouth, indicative of breathlessness; Raj and Gregory, 1996) and vocalisations (Holst, 2001; Velarde et al., 2007).

The main hazards causing increased distress during induction to unconsciousness are irritant or aversive gas mixtures, low gas temperature or humidity.

The exposure to anoxic gases (argon or nitrogen) with less than 2% by volume of residual oxygen in air is non-aversive and does not cause sense of breathlessness before the loss of consciousness occurs. However, the time to induce unconsciousness when exposed to anoxia is longer than when exposed to hypercapnia (Raj et al., 1997) and might not be commercially feasible. Raj and Gregory (1996) reported that the addition of up to 30% CO_2 to an anoxic atmosphere reduces the time needed to induce unconsciousness with minimal aversion. Argon (Ar) has a low presence in the atmosphere (0.9% by volume), and its availability for commercial stunning practices might be limited. On the other hand, the relative density of nitrogen (N_2) is slightly lower than air concentrations and cannot be sustained within a pit at a higher concentration than 94% by volume (Dalmau et al., 2010). Nevertheless, this stability could be improved when nitrogen and CO_2 are combined (Dalmau et al., 2010).

The EU Regulation (EC 1099/2009) has approved the low atmospheric pressure stunning (LAPS) for broiler chickens. It consists of the exposure to gradual decompression with reduction in the available oxygen to less than 5% (Martin et al., 2016). Current research is carried out in pigs to assess the effectiveness in inducing unconsciousness without aversion.

An effective stun is characterised by the presence of the loss of posture, apnoea, absence of corneal reflex and absence of muscle tone (Velarde et al., 2007). The main hazards causing ineffective stunning are incorrect gas concentration or short gas exposure time (EFSA, 2004). In case of ineffective stunning or recovery, re-stun immediately using a backup system taking into account the causes of failure or recovery.

Comparative studies reported the positive effects of CO_2 stunning on the pork quality when compared with electrical stunning (Velarde et al., 2000, 2001). However, the effects of gas stunning on animal welfare and pork quality depend on the gas concentration and exposure time. Better results, in terms of the percentage of clinical reflexes and risk for PSE pork, were obtained after stunning with a higher CO_2 concentration (90% vs. 80%) and longer gas exposure time (100 vs. 70 sec; Nowak et al., 2007).

4.3 Exsanguination techniques

Exsanguination consists of the severance of the arteries supplying oxygenated blood to the brain. Pigs are bled by chest sticking, severing the brachiocephalic trunk and the major blood vessels which arise from the heart (Fig. 8). With an adequate incision, pigs lose between 40% and 60% of their total blood volume, and within 30 sec, the amount of blood lost is between 70% and 80% of the total amount of blood which will be lost (Warriss and Wilkins, 1987). The time to lose brain responsiveness (based on the reduction in visual evoked responses) ranges between 14 sec and 23 sec (mean 18 ± 63) and the development of an isoelectric electrocorticogram (ECoG) between 22 sec and 30 sec (Wotton and Gregory, 1986).

The main welfare concern at the time of bleeding following stunning is the recovery of consciousness due to prolonged stun-to-stick interval or due to incomplete severance of the main blood vessels.

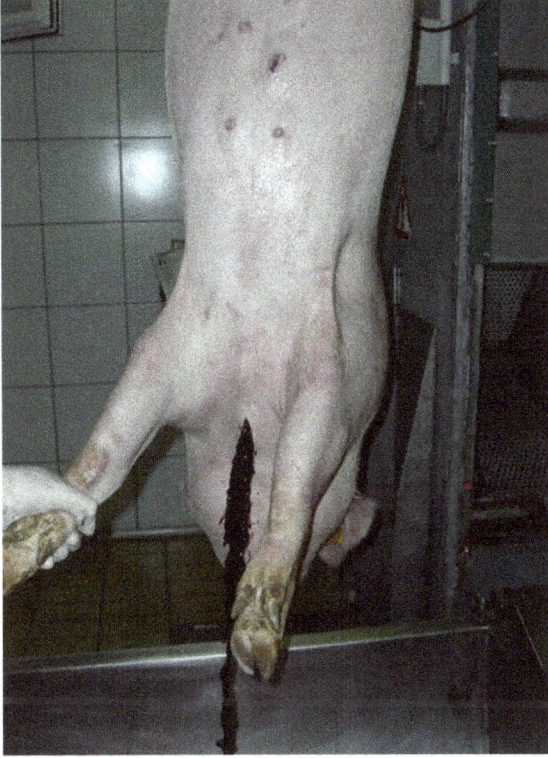

Figure 8 Chest sticking (A. Velarde, IRTA, Spain).

5 Animal welfare audit protocols

Nowadays, there is an increasing need for credible assessment systems to determine the welfare status of animals prior to slaughter. The application of animal welfare auditing protocols allows the evaluation of handling and slaughter practices, resulting in a significant improvement in handling practices, facilities design and quality of work, besides reassuring concerned consumers and increasing market opportunities due to the availability of certified products (Ballantyne, 2006; Grandin, 2017a).

Because of these benefits, animal welfare auditing has been increasingly implemented during transport and at slaughter since 1999 in the US, Australia, New Zealand and Europe, and in Canada since 2001 (Grandin and Smith, 2004).

Presently, the two main protocols used to assess animal welfare during transport and lairage and at slaughter are the audit protocols developed by Temple Grandin for the North American Meat Institute (NAMI; Grandin, 2017b) and by researchers of the European project Welfare Quality® (WQ®, 2009).

The NAMI audit protocol is based on the 'Recommended animal handling guidelines and audit guide' and has the objective to evaluate animal welfare during transport and at the slaughter plant through the assessment of seven core criteria based on the observation of animal behaviour (falls and vocalisation), assessment of facilities design, electric prod use, wilful acts of abuse, efficiency of stunning systems (electrical and CO_2 gas), access to water and insensibility on the bleed rail (Grandin, 2017b). The respect of the NAMI guidelines at the slaughter plant is a requirement to comply with the standards of the Humane Farm Animal Care and for the pork product to get the Certified Humane Raised and Handled® label (Humane Farm Animal Care, 2018).

The WQ® protocol (WQ®, 2009) has developed an integrated and standardised welfare assessment system based on 12 welfare criteria grouped into four main principles (good feeding, good housing, good health and appropriate behaviour) according to how they are experienced by animals (Dalmau et al., 2009). The innovation of the WQ® assessment system is its greater focus on animal-based measures (e.g. body condition, health, injuries and behaviour) rather than on resource and management-based measures (e.g. space allowance, number of drinkers and truck management). The WQ® welfare assessment starts in the unloading area, where general fear, thermoregulation behaviours, slipping and falling, sickness and dead animals are observed (Velarde and Dalmau, 2012). In lairage, five criteria are taken into consideration, namely the absence of thirst, hunger and disease; thermal comfort; and comfort while resting. The absence of thirst is calculated based on the number of drinking points in each pen and drinker functionality and cleanliness. The availability of feed for animals that have been held more than 12 h in the holding pen is the measure for the absence of hunger, while the absence of disease is measured by the presence of dead animals in the pen. Thermal comfort is evaluated by behavioural thermoregulation measures, such as huddling, shivering or panting scores. The last, but not least, criterion is comfort while resting in the pen, which is evaluated based on the space allowance. During movement from the lairage pen to the stunning area, good human–animal relationship is assessed by recording the incidence of high-pitched vocalisations. Afterwards, stunning effectiveness is assessed by observing the signs of the absence of pain, such as corneal and righting reflex, breathing and vocalisation, immediately after stunning and before sticking. After slaughter, skin lesion scoring provides valuable information regarding the management of animals on the farm of origin, during transport or in the lairage pen. Furthermore, the health status of the animals on the farm of origin is also assessed after slaughter by inspecting the internal organs for the presence of pleurisy and pneumonia in the lungs, pericarditis in the heart and white spots in the liver.

The efficiency of the WQ® programme as a tool for animal welfare monitoring has been scientifically validated at the abattoir (Dalmau et al., 2009). In 2014, a certification scheme of animal welfare based on the WQ® protocols has been developed by the Animal Welfare Subprogram of the IRTA (Monells, Spain) and the Spanish Association for Standardisation and Certification (AENOR). In 2017, already several pig abattoirs had been certified. The final objective of this programme is to include the certificate in the pork meat product label, named as 'Animal Welfare. AENOR Conform' (Dalmau, 2017). Recently, the IRTA has become scheme owner for the Animal Welfare Certification 'based on Welfare Quality®'.

However, neither the WQ® nor the NAMI assessment protocols cover the production process as a whole (from farm to slaughter). Rocha et al. (2016) developed a novel protocol for the animal welfare assessment of the whole pork chain by merging the WQ protocol with those of the Canadian Animal Care Assessment (CPC, 2011), which is a more resource- and management-based measurement of animal welfare at the farm level (including loading), and that of the American Meat Institute (AMI; Grandin, 2012) and validated it under Canadian commercial conditions. The objective of this study was to assess the relationship between the audit scores obtained at animal welfare-improved (AWI) and conventional farms using the WQ and CPC protocols and the variation in pig behaviours at loading, at unloading and in lairage as assessed by the WQ and AMI audit protocols. In this study, the frequency of slips used as an animal welfare audit criterion in the AMI protocol at the slaughter plant showed to be a good predictor of drip loss variation in the loin muscle, with muscle exudation being significantly related to slips during unloading ($r = 0.63$) and immediately before stunning ($r = 0.74$).

6 Conclusion and future trends

This chapter overviewed the effects of the stressors experienced by pigs during the transport, lairage and slaughter on animal losses; behavioural and physiological responses to stress; and carcass and meat quality.

Research provided the evidence that the use of truck models featuring hydraulic ramps or decks can help reduce the workload of handlers and improve the welfare of pigs during transport. However, more research on the truck design is needed with a study of the air flow patterns and vibration rate and insulation and cooling systems under different ambient conditions where temperature control becomes more critical and physiological heat maintenance and dissipation in pigs of different market weights become less effective. So far, most swine transportation research focused on slaughter weight pigs (up to 130 kg), but there is a considerable lack of information for the transportation of heavier pigs, including cull sows and boars. More specifically, information on their thermal and physical needs during transport and relative science-based guidelines for space allowance and travel time are needed.

Lairage and slaughter are extremely important for the pork chain economy as mistakes made at these points have irreversible effects on the welfare of pigs and carcass and meat quality and may offset all efforts made by the production sector to improve performance and animal welfare. Precautions must be taken to ensure proper handling and ambient control to keep the benefits of lairage as a resting area, allowing pigs to recover from the stress of transport and previous handling. The correct management and monitoring of critical areas in lairage is becoming paramount in the light of the increasing need for commercial abattoirs to obtain animal welfare audit approval and certification for their meat products.

7 Where to look for further information

7.1 Further reading

- Comprehensive and up-to-date reviews on swine transportation and animal welfare are the chapters authored by Faucitano and Goumon (2018) and Faucitano and Lambooij (2019) and published in the books Advances in Pig Welfare and Livestock Transport and Handling.
- An excellent source of science-based guidelines for the best handling practices at the abattoir (from reception to slaughter) in multiple species is the book Animal Welfare at Slaughter: a Practitioner Guide by Raj and Velarde (2016).
- Guidelines and recommendations on preslaughter handling and animal welfare can be found in the Temple Grandin web page: http://www.grandin.com/.
- A complete introduction to the relationship between animal welfare preslaughter and meat science is represented by the books Animal Welfare and Meat Science (1998) and Animal Welfare and Meat Production (2007) by Neville Gregory.
- A catalogue of training handbooks, guidelines and technical sheets in relation to transport and slaughter are available in the web page of the Humane Slaughter Association: https://www.hsa.org.uk/.

7.2 Key journals/conferences

- Special issues on pig transportation in the Animals journal (2014/2016): https://www.mdpi.com/journal/animals/special_issues/pig-trans-2016; https://www.mdpi.com/journal/animals/special_issues/pig-trans.
- Proceedings of special sessions on preslaughter handling and animal welfare included in the programme of the annual congresses of the

International Society of Applied Ethology (ISAE; https://www.applied-ethology.org/), the International Congress of Meat Science & Technology (ICoMST; http://www.icomst.helsinki.fi/index.htm) and American and Canadian Societies of Animal Science (ASAS-CSAS) Meeting & Trade Show; https://www.asas.org).

7.3 Major international research projects

- The Canadian Agricultural Partnership (http://www.agr.gc.ca/eng/about-us/key-departmental-initiatives/canadian-agricultural-partnership/?id=1461767369849) and Swine Innovation Porc (http://www.swineinnovationporc.ca/) are supporting projects on swine transportation.

8 References

Anil, M. H. (1991), 'Studies on the return of physical reflexes in pigs following electrical stunning', *Meat Sci.*, 30, 13–21.

Averós, X., Knowles, T. G., Brown, S. N., Warriss, P. D. and Gosálvez, L. F. (2008), 'Factors affecting the mortality of pigs being transported to slaughter', *Vet. Rec.*, 163, 386–90.

Ballantyne, W. (2006), 'A proactive approach to animal welfare', *Adv. Pork Prod.*, 17, 139–44.

Baloyh, P., Kapelankis, W., Jankowiak, H., Nagy, L., Kovacs, S., et al. (2015), 'The productive lifetime of sows on two farms and reasons for culling', *Ann. Anim. Sci.*, 15, 747–58.

Barton-Gade, P. and Christensen, L. (1998), 'Effect of different loading density during transport on welfare and meat quality in Danish slaughter pigs', *Meat Sci.*, 48, 237–47.

Barton-Gade, P., Christensen, L., Brown, S. N. and Warriss, P. D. (1996), 'Effect of tier and ventilation during transport on blood parameters and meat quality in slaughter pigs', *EU-Seminar: New Information on Welfare and Meat Quality of Pigs as Related to Handling, Transport and Lairage Conditions*, Landbauforschung Völkenrode, Kulmbach, Germany, 1996, vol. 166, 101–16.

Barton-Gade, P., Christensen, L., Baltzer, M. and Petersen, L. (2007), 'Causes of pre-slaughter mortality in Danish slaughter pigs with special emphasis on transport', *Anim. Welf.*, 16, 459–70.

Becerril-Herrera, M. M., Alonso-Spilsbury, M., Trujillo-Ortega, M. E., Guerrero-Legarreta, I., Ramirez-Necoechea, R., et al. (2010), 'Changes in blood constituents of swine transported for 8 or 16 h to an abattoir', *Meat Sci.*, 86, 945–8.

Bench, C., Schaefer, A. L. and Faucitano, L. (2008), 'The welfare of pigs during transport', in L. Faucitano and A. L. Schaefer (Eds), *Welfare of Pigs: From Birth to Slaughter*, Wageningen, The Netherlands: Wageningen Academic Publishers, pp. 161–95.

Bertol, T. M., Ellis, M., Ritter, M. J. and McKeith, F. K. (2011), 'Effect of feed withdrawal and handling intensity on longissimus muscle glycolytic potential and blood measurements in slaughter weight pigs', *J. Anim. Sci.*, 83, 1536–42.

Blair, B. and Lowe, J. (2019), 'Describing the cull sow market network in the US: a pilot project', *Prev. Vet. Med.*, 162, 107–9.

Bottaccini, M., Scollo, A., Edwards, S., Contiero, B., Veloci, M., et al. (2018), 'Skin lesion monitoring at slaughter on heavy pigs (170 kg): welfare indicators and ham defects', *PLoS ONE* (open access). doi: 10.1371/journal.pone.0207115.

Brent, G. (1986), *Housing the Pig*, Ipswich, UK: Farming Press Ltd.

Brown, S. N., Knowles, T. G., Edwards, J. E. and Warriss, P. D. (1999), 'Relationship between food deprivation before transport and aggression in pigs held in lairage before slaughter', *Vet. Rec.*, 145, 630–34.

Brown, J., Samarakone, T. S., Crowe, T., Bergeron, R., Widowski, T. M., et al. (2011), 'Temperature and humidity conditions in trucks transporting pigs in two seasons in eastern and western Canada', *Trans. ASABE*, 54, 1–8.

CARC. (2001), 'Recommended code of practice for the care and handling of farm animals – transportation', Canadian Agri-Food Research Council, Ottawa, Canada, 63p.

Chevillon, P. (2001), 'Pig welfare during pre-slaughter and stunning', *1st International Virtual Conference on Pork Quality*, Brazil, pp. 145–58.

Christensen, L. and Barton-Gade, P. (1997), 'New Danish developments in pig handling at abattoirs', *Fleischwirt.*, 77, 604–7.

Conte, S., Faucitano, L., Bergeron, R., Torrey, S., Gonyou, H. W., et al. (2015), 'Effects of season, truck type and location within truck on gastrointestinal tract temperature of market-weight pigs during transport', *J. Anim. Sci.*, 93, 5840–8.

Correa, J. A. (2011), 'Effect of farm handling and transport on physiological response, losses and meat quality of commercial pigs', *Adv. Pork Prod.*, 22, 249–56.

Correa, J. A., Torrey, S., Devillers, N., Laforest, J. P., Gonyou, H. W. and Faucitano, L. (2010), 'Effects of different moving devices at loading on stress response and meat quality in pigs', *J. Anim. Sci.*, 88, 4086–93.

Correa, J. A., Gonyou, H. W., Torrey, S., Widowski, T., Bergeron, R., et al. (2013), 'Welfare and carcass and meat quality of pigs being transported for 2 hours using two vehicle types during two seasons of the year', *Can. J. Anim. Sci.*, 93, 43–55

Correa, J. A., Gonyou, H. W., Torrey, S., Widowski, T., Bergeron, R., et al. (2014), 'Welfare of pigs being transported for long distance using a pot-belly trailer during winter and summer', *Animals*, 4, 200–13.

Council Regulation (EC) 1099/2009 (2009), 'Council Regulation No 1099/2009 on the protection of animals at the time of killing', *OJEU*, L303, 1–30.

CPC (2011), 'Animal care assessment', Canadian Pork Council, Ottawa, Canada. http://www.cqa-aqc.com/aca/documents/ACA-Animal-Care-Assessment.pdf.

Croft, P. G. (1952), 'Problem of electrical stunning', *Vet. Rec.*, 64, 255–8.

Dalla Costa, O. A., Faucitano, L., Coldebella, A., Ludke, J. V., Peloso, J. V., et al. (2007), 'Effects of the season of the year, truck type and location on truck on skin bruises and meat quality in pigs', *Livest. Sci.*, 107, 29–36.

Dalla Costa, F. A., Devillers, N., Paranhos da Costa, M. J. R. and Faucitano, L. (2016), 'Effects of applying preslaughter feed withdrawal at the abattoir on behaviour, blood parameters and meat quality in pigs', *Meat Sci.*, 119, 89–94.

Dalla Costa, O. A., Dalla Costa, F. A., Feddern, V., Dos Santos Lopes, L., et al. (2019), 'Risk factors associated with pig pre-slaughtering losses', *Meat Sci.*, 155, 61–8.

Dalmau, A. (2017), 'Development of a certification schema on animal welfare based on Welfare Quality protocols', *Newsl. Welf. Qual. Netw.*, 5, 2–3.

Dalmau, A., Temple, D., Rodríguez, P., Llonch, P. and Velarde, A. (2009), 'Application of the Welfare Quality® protocol at pig slaughterhouses', *Anim. Welf.*, 18, 497–505.

Dalmau, A., Llonch, P., Rodríguez, P., Ruíz-de-la-Torre, J. L., et al. (2010), 'Stunning pigs with different gas mixtures. Part 1: gas stability', *Anim. Welf.*, 19, 315–23.

Dewey, C., Haley, C., Widowski, T., Poljak, Z. and Friendship, R. (2009), 'Factors associated with in-transit losses of fattening pigs', *Anim. Welf.*, 18, 355–61.

Dokmanović, M., Velarde, A., Tomović,V., Glamočlija, N., Marković, R., et al. (2014), 'The effects of lairage time and handling procedure prior to slaughter on stress and meat quality parameters in pigs', *Meat Sci.*, 98, 220–6.

Edwards, L. N., Grandin, T. A., Engle, T. E., Porter, S. P., Ritter, M. J., et al. (2010), 'Use of exsanguination blood lactate to assess the quality of preslaughter pig handling', *Meat Sci.*, 86, 384–90.

Edwards, L. N, Engle, T. E., Grandin, T., Ritter, M. J., Sosnicki, A., et al. (2011), 'The effects of distance traveled during loading, lairage time prior to slaughter, and distance traveled to the stunning area on blood lactate concentration of pigs in a commercial packing plant', *Prof. Anim. Sci.*, 27, 485–91.

EFSA (2004), 'Welfare aspects of animal stunning and killing methods', Scientific Report of the Scientific Panel for Animal Health and Welfare on request from the Commission related to welfare aspects of animal stunning and killing methods (Question No. EFSA-Q-2003-093). European Food Safety Authority (AHAW 04-027).

EFSA (2013), 'Scientific Opinion on monitoring procedures at slaughterhouses for pigs', *EFSA J.*, 11(12), 3523.

Eyes on Animals (2019), 'Improving animal welfare in pig slaughterhouses'. http://www .eyesonanimals.com/wp-content/uploads/2016/06/Animal-welfare-in-pig-slaught erhouses-how-to-reduce-stress-suffering-and-ease-handling-aanp-1.pdf (accessed on 14 May 2019).

Faucitano, L. (2001), 'Causes of skin damage to pig carcasses', *Can. J. Anim. Sci.*, 81, 39–45.

Faucitano, L. (2010), 'Effects of lairage and slaughter conditions on animal welfare and pork quality', *Can. J. Anim. Sci.*, 90, 461–9.

Faucitano, L. (2018), 'Preslaughter handling practices and their effects on animal welfare and pork quality', *J. Anim. Sci.*, 96, 728–38.

Faucitano, L. and Goumon, S. (2018), 'Transport to slaughter and associated handling', in M. Špinka (Ed.), *Advances in Pig Welfare*, London, UK: Woodhead Publishing, pp. 261–93.

Faucitano, L. and Lambooij, E. (2019), 'Transport of pigs', in T. Grandin (Ed.), *Livestock Transport and Handling* (5th edn.), Wallingford, UK: CABI Publishing, pp. 302–27.

Fitzgerald, R. F., Stalder, K. J., Matthews, J. O., Schultz-Kaster, C. M. and Johnson, A. K. (2009), 'Factors associated with fatigued, injured, and dead pig frequency during transport and lairage at a commercial abattoir', *J. Anim. Sci.*, 87, 1156–66.

Fox, J., Widowski, T., Torrey, S., Nannoni, E., Bergeron, R., et al. (2014), 'Water sprinkling market pigs in a stationary trailer. 1. Effects on pig behaviour, gastrointestinal tract temperature and trailer micro-climate', *Livest. Sci.*, 160, 113–23.

Fraqueza, M. J., Roseiro, L. C., Almeida, J., Matias, E., Santos, C., et al. (1998), 'Effects of lairage temperature and holding time on pig behavior and on carcass and meat quality', *Appl. Anim. Behav. Sci.*, 60, 317–30.

Gallo, C., Faucitano, L. and Gerritzen, M. (2016), 'Effects of preslaughter handling on carcass and meat quality', in M. Raj and A. Velarde (Eds), *Animal Welfare at Slaughter: a Practitioner Guide*, Sheffield, UK: 5m Publishing, pp. 251–69.

Garcia, A. and McGlone, J. J. (2015), 'Loading and unloading finishing pigs: effects of bedding types, ramp angle, and bedding moisture', *Animals*, 5, 13–26.

Gerritzen, M. A., Hindle, V. A., Steinkamp, K. and Reimert, H. G. M. (2013), 'The effect of reduced loading density on pig welfare during long distance transport', *Anim. 7*, 1849-57.

Geverink, N. A., Engel, B., Lambooij, E. and Wiegant, V. M. (1996), 'Observations on behavior and skin damage of slaughter pigs and treatment during lairage', *Appl. Anim. Behav. Sci.*, 50, 1-13.

Geverink, N. A., Buhnemann, A., Van de Burgwal, J. A., Lambooij, E., Blokhuis, H. J., et al. (1998), 'Responses of slaughter pigs to transport and lairage sounds', *Physiol. Behav.*, 63, 667-73.

Gonyou, H. W. and Brown, J. (2012), 'Reducing stress and improving recovery from handling during loading and transport of market pigs', Final Report submitted to Alberta Livestock and Meat Agency, Edmonton, Canada, 40p.

Goumon, S. and Faucitano, L. (2017), 'Influence of loading handling and facilities on the subsequent response to pre-slaughter stress in pigs', *Livest. Sci.*, 200, 6-13.

Goumon, S., Brown, J. A., Faucitano, L., Bergeron, R., Widowski, T. M., et al. (2013a), 'Effects of transport duration on maintenance behavior, heart rate and gastrointestinal tract temperature of market-weight pigs in 2 seasons', *J. Anim. Sci.*, 91, 4925-35.

Goumon, S., Faucitano, L., Bergeron, R., Crowe, T., Connor, M. L., et al. (2013b), 'Effect of ramp configuration on easiness of handling, heart rate and behavior of near-market pigs at unloading', *J. Anim. Sci.*, 91, 3889-98.

Grandin, T. (1990), 'Design of loading facilities and holding pens', *Appl. Anim. Behav. Sci.*, 28, 187-201.

Grandin, T. (2010), 'The importance of measurement to improve the welfare of livestock, poultry and fish', in T. Grandin (Ed.), *Improving Animal Welfare: a Practical Approach*, Wallingford, UK: CABI Publishing, pp. 1-20.

Grandin, T. (2012), *Recommended Animal Handling Guidelines and Audit Guide: a Systematic Approach to Animal Welfare*, Washington, DC: American Meat Institute Foundation, 108p.

Grandin, T. (2016a), 'Transport fitness of cull sows and boars: a comparison of different guidelines of fitness for transport', *Animals*, 6, 77 (open access). doi: 10.3390/ani8070124.

Grandin, T. (2016b), 'Practical methods to improve animal handling and restraint', *in* A. Velarde and M. Raj (Eds), *Animal Welfare at Slaughter*, Sheffield, UK: 5M Publishing, pp. 71-90.

Grandin, T. (2017a), 'How to work with large meat buyers to improve animal welfare', *in* P. P. Purslow (Ed.), *New Aspects of Meat Quality: from Genes to Ethics*, London, UK: Woodhead Publishing, pp. 569-79.

Grandin, T. (2017b), *Recommended Animal Handling Guidelines and Audit Guide: Systematic Approach to Animal Welfare*, Washington DC: North American Meat Institute Foundation, 108p.

Grandin, T. (2018), 'Welfare problems in cattle, pigs, and sheep that persist even though scientific research clearly shows how to prevent them', *Animals*, 8, 124 (open access). doi: 10.3390/ani8070124.

Grandin, T. and Smith, G. C. (2004), 'Animal welfare and humane slaughter'. http://www.grandin.com/references/humane.slaughter.html (accessed on 23 September 2019).

Griot, B., Boulard, J., Chevillon, P. and Kerisit, R. (2000), 'Des restrainers à bande pour le bienêtre et la qualité de la viande', *Viandes et Produits Carnés*, 3, 91-7.

Guàrdia, M. D., Gispert, M. and Diestre, A. (1996), 'Mortality rates during transport and lairage in pigs for slaughter', *Meat Focus Inter.*, 10, 362–6.

Guàrdia, M. D., Estany, J., Balasch, S., Oliver, M. A., Gispert, M., et al. (2004), 'Risk assessment of PSE condition due to pre-slaughter conditions and RYR1 gene in pigs', *Meat Sci.*, 67, 471–8.

Guàrdia, M. D, Estany, J., Balasch, S., Oliver, M. A., Gispert, M., et al. (2005), 'Risk assessment of DFD meat due to pre-slaughter conditions in pigs', *Meat Sci.*, 70, 709–16.

Guàrdia, M. D., Estany, J., Balasch, S., Oliver, M. A., Gispert, M., et al. (2009), 'Risk assessment of skin damage due to pre-slaughter conditions and RYR1 gene in pigs', *Meat Sci.*, 81, 745–51.

Haley, C., Dewey, C. E., Widowski, T. and Friendship, R. (2008), 'Association between in-transit losses, internal trailer temperature, and distance travelled by Ontario market hogs', *Can. J. Vet. Res.*, 72, 385–9.

Haley, C., Dewey, C. E., Widowski, T. and Friendship, R. (2010), 'Relationship between estimated finishing-pig space allowance and in transit loss in a retrospective survey for 3 packing plant in Ontario in 2003', *Can. J. Vet. Res.*, 74, 178–84.

Hambrecht, E., Eissen, J. J., Newman, D. J., Smits, C. H., Verstegen, M. W., et al. (2005), 'Preslaughter handling effects on pork quality and glycolytic potential in two muscles differing in fiber type composition', *J. Anim. Sci.*, 83, 900–7.

Hoenderken, R. (1978), 'Electrical stunning of pigs for slaughter', Doctoral thesis, University of Utrecht, NL.

Holst, S. (2001), 'Carbon dioxide stunning of pigs for slaughter – practical guidelines for good animal welfare', *47th International Congress of Meat Science and Technology*, Krakow, Poland, pp. 48–54.

Honkavaara, M. (1989), 'Influence of lairage on blood composition of pig and on the development of PSE pork', *J. Agric. Sci. Finland*, 61, 433–40.

Humane Farm Animal Care (2018), 'Humane farm animal care standards', Humane Farm Animal Care, Middleburg, VA, 31p. www.certifiedhumane.org.

Huynh, T. T. T., Aarnink, A. J. A., Truong, C. T., Kemp, B. and Verstegen, M. W. A. (2006), 'Effects of tropical climate and water cooling on growing pigs' responses', *Livest. Sci.*, 104, 278–91.

Kephart, K. B., Harper, M. T. and Raines, C. R. (2010), 'Observations of market pigs following transport to a packing plant', *J. Anim. Sci.*, 88, 2199–203.

Knowles, T. G., Brown, S. N., Edwards, J. E. and Warriss, P. D. (1998), 'Ambient temperature below which pigs should not be continuously showered in lairage', *Vet. Rec.*, 143, 575–8.

Koketsu Y. (2000), 'Factors associated with increased sow mortality in North America', *Proceedings of the American Association of Swine Practitioners*, pp. 419–20.

Lambooij, E. (1988), 'Road transport of pigs over a long distance: some aspects of behaviour, temperature and humidity during transport and some effects of the last two factors', *Anim. Prod.*, 46, 257–63.

Lambooij, E. (2014), 'Transport of pigs', *in* T. Grandin (Ed.), *Livestock Handling and Transport*, Wallingford, UK: CABI Publishing, pp. 280–97.

Lambooij, E. and Engel, B. (1991), 'Transport of slaughter pigs by road over a long distance: some aspects of loading density and ventilation', *Livest. Prod. Sci.*, 28, 163–74.

Lambooij, E., Garssen, G. J., Walstra, P., Mateman, F. and Merkus, G. S. M. (1985), 'Transport of pigs by car for two days: some aspects of watering and loading density', *Livest. Prod. Sci.*, 13, 289–99.

Leheska, J. M., Wulf, D. M. and Maddock, R. J. (2003), 'Effects of fasting and transportation on pork quality development and extent of postmortem metabolism', *J. Anim. Sci.*, 80, 3194-202.

Long, V. P. and Tarrant, P. V. (1990), 'The effect of pre-slaughter showering and post-slaughter rapid chilling on meat quality in intact pork sides', *Meat Sci.*, 27, 181-95.

Malena, M., Voslárová, E., Kozák, A., Belobrádek, P., Bedánová, I., et al. (2007), 'Comparison of mortality rates in different categories of pigs and cattle during transport for slaughter', *Acta Vet. Brno*, 76, 109-16.

Martin, J. E., Christensen, K., Vizzier-Thaxton, Y. and McKeegan, D. E. F. (2016), 'Effects of analgesic intervention on behavioural responses to Low Atmospheric Pressure Stunning', *Appl. Anim. Behav. Sci.*, 180, 157-65.

McGee, M., Johnson, A. K., O'Connor, A. M., Tapper, K. R. and Millman, S. T. (2016), 'An assessment of swine marketed through buying stations and development of fitness for transport guidelines', *J. Anim. Sci.*, 94, 9.

McGlone, J. J., Johnson, A. K., Sapkota, A. and Kephart, R. K. (2014a), 'Establishing trailer ventilation (boarding) requirements for finishing pigs during transport', *Animals*, 4, 515-23.

McGlone, J. J., Johnson, A. K., Sapkota, A. and Kephart, R. K. (2014b), 'Transport of market pigs: improvements in welfare and economics', *in* T. Grandin (Ed.), *Livestock Handling and Transport*, Wallingford, UK: CABI Publishing, pp. 298-314.

Moss, B. W. (1978), 'Some observations on the activity and aggressive behavior of pigs when penned prior to slaughter', *Appl. Anim. Ethol.*, 4, 323-39.

Mota-Rojas, D., Becerril, M., Lemus, C., Sánchez, P., González, M., Olmos, S. A., et al. (2006), 'Effects of mid-summer transport duration on pre- and post-slaughter performance and pork quality in Mexico', *Meat Sci.*, 73, 404-12.

Nanni Costa, L., Lo Fiego, D. P., Dall'Olio, S., Davoli, R. and Russo, V. (2002), 'Combined effects of pre-slaughter treatments and lairage time on carcass and meat quality in pigs of different halothane genotype', *Meat Sci.*, 61, 41-7.

Nannoni, E., Widowski, T. M., Torrey, S., Fox, J., Rocha, L. M., et al. (2014), 'Water sprinkling market pigs in a stationary trailer. 2. Effects on selected exsanguination blood parameters and carcass and meat quality variation', *Livest. Sci.*, 160, 124-31.

Nannoni, E., Liuzzo, G., Serraino, A., Giacometti, F., Martelli, G., et al. (2016), 'Evaluation of pre-slaughter losses of Italian heavy pigs', *Anim. Prod. Sci.* (open-access). doi: 10.1071/AN15893.

Nowak, B., Mueffling, T. V. and Hartung, J. (2007), 'Effect of different carbon dioxide concentrations and exposure times in stunning of slaughter pigs: impact on animal welfare and meat quality', *Meat Sci.* 75, 290-8.

Pereira, T., Titto, E. A., Conte, S., Devillers, N., Sommavilla, R., et al. (2018), 'Use of fan-misters bank for cooling pigs kept in a stationary trailer before unloading: effects on trailer microclimate, and pig behavior and physiological response', *Livest. Sci.*, 216, 67-74.

Peterson, E., Remmenga, M., Hagerman, A. D. and Akkina, J. E. (2017), 'Use of temperature, humidity, and slaughter condemnation data to predict increases in transport losses in three classes of swine and resulting foregone revenue', *Front. Vet. Sci.*, 4, 67 (open access). doi: 10.3389/fvets.2017.00067.

Pilcher, C. M., Ellis, M., Rojo-Gomez, A., Curtis, S. E., Wolter, B. F., et al. (2011), 'Effects of floor space during transport and journey time on indicators of stress and transport losses in market weight pigs', *J. Anim. Sci.*, 89, 3809-18.

Rabaste, C., Faucitano, L., Saucier, L., Foury, D., Mormède, P., et al. (2007), 'The effects of handling and group size on welfare of pigs in lairage and its influence on stomach weight, carcass microbial contamination and meat quality variation', *Can. J. Anim. Sci.*, 87, 3-12.

Raj, A. B. M. and Gregory, N. G. (1996), 'Welfare implications of the gas stunning of pigs. Stress of induction of anaesthesia', *Anim. Welf.*, 5, 71-8.

Raj, A. B. M. and Velarde, A. (2016), *Animal Welfare at Slaughter: A Practitioner Guide*, Sheffield, UK: 5m Publishing.

Raj, A. B. M., Johnson, S. P., Wotton, S. B. and McInstry, J. L. (1997), 'Welfare implications of gas stunning of pigs 3. Time to loss of somatosensory evoked potentials and spontaneous electrocorticogram of pigs during exposure to gases', *Br. Vet. J.*, 153, 329-40.

Renaudeau, D., Gourdine, J. and St-Pierre, N. (2011), 'A meta-analysis of the effects of high ambient temperature on growth performance of growing-finishing pigs', *J. Anim. Sci.*, 89, 2220-30.

Rioja-Lang, F. C., Brown, J. A., Brockhoff, E. J. and Faucitano, L. (2019), 'A review of swine transportation research on priority welfare issues: a Canadian perspective', *Front. Vet. Sci.*, 6, 36 (open access). doi: 10.3389/fvets.2019.00036.

Ritter, M. J., Ellis, M., Brinkmann, J., DeDecker, J. M., Keffaber, K. K., et al. (2006), 'Effect of floor space during transport of market-weight pigs on the incidence of transport losses at the packing plant and the relationships between transport conditions and losses', *J. Anim. Sci.*, 84, 2856-64.

Ritter, M. J., Ellis, M., Bowman, R., Brinkmann, J., Curtis, S. E., et al. (2008), 'Effects of season and distance moved during loading on transport losses of market-weight pigs in two commercially available types of trailer', *J. Anim. Sci.*, 86, 3137-45.

Ritter, M., Rincker, P. and Carr, S. (2012), 'Pig handling and transportation strategies utilized under U.S. commercial conditions', London Swine Conference, London, Canada, pp. 109-20.

Rocha, L. M., Bridi, A. M., Foury, A., Mormède, P., Weschenfelder, A. V., et al. (2013), 'Effects of ractopamine administration and castration method on the response to pre-slaughter stress and carcass and meat quality in pigs of two Piétrain genotypes', *J. Anim. Sci.*, 91, 3965-77.

Rocha, L. M., Dionne, A., Saucier, L., Nannoni, E. and Faucitano, L. (2015), 'Hand-held lactate analyzer as a tool for the real-time measurement of physical fatigue before slaughter and pork quality prediction', *Anim.* 9, 707-14.

Rocha, L. M., Velarde, A., Dalmau, A., Saucier, L. and Faucitano, L. (2016), 'Can the monitoring of animal welfare parameters predict pork meat quality variation through the supply chain (from farm to slaughter)?', *J. Anim. Sci.*, 94, 359-76.

Rodríguez, P., Dalmau, A., Ruiz-de-la-Torre, J. L., Manteca, X., Jensen, E. W., et al. (2008), 'Assessment of unconsciousness during carbon dioxide stunning in pigs', *Anim. Welf.*, 17, 341-9.

Rydhmer, L., Zamaratskaia, G., Andersson, H. K., Algers, B., Guillemet, R., et al. (2006), 'Aggressive and sexual behavior of growing and finishing pigs reared in groups, without castration', *Acta Agric. Scand., Sect. A Anim. Sci.*, 56, 109-19.

Santos, C., Almeida, J. M., Matias, E. C., Fraqueza, M. J., et al. (1997), 'Influence of lairage environmental conditions and resting time on meat quality in pigs', *Meat Sci.*, 45, 253-62

SCAHAW (Scientific Committee on Animal Health and Animal Welfare) (2011), 'Scientific opinion concerning the welfare of animals during transport', *EFSA J.*, 9, 1966.

Scheeren, M. B., Gonyou, H. W., Brown, J., Weschenfelder, A. V. and Faucitano, L. (2014), 'Effects of transport time and location within truck on skin bruises and meat quality of market weight pigs in two seasons', *Can. J. Anim. Sci.*, 94, 71-8.

Schwartzkopf-Genswein, K. S., Faucitano, L., Dadgar, S., Shand, P., Gonzàlez, L. A., et al. (2012), 'Road transportation of cattle, swine and poultry in North America and its impact on animal welfare, carcass and meat quality: a review', *Meat Sci.*, 92, 227-43.

Shen, Q. W., Means, W. J., Thompson, S. A., Underwood, K. R., Zhu, M. J., et al. (2006), 'Pre-slaughter transport, AMP-activated protein kinase, glycolysis, and quality of pork loin', *Meat Sci.*, 74, 388-95.

Sommavilla, R., Faucitano, L., Gonyou, H. W., Seddon, Y., Bergeron, R., et al. (2017), 'Season, transport duration and trailer compartment effects on blood stress indicators in pigs: relationship to environmental, behavioural and other physiological factors, and pork quality traits', *Animals*, 7, 8 (open access). doi: 10.3390/ani7020008.

Sutherland, M. A., McDonald, A. and McGlone, J. J. (2009), 'Effects of variations in the environment, length of journey and type of trailer on the mortality and morbidity of pigs being transported to slaughter', *Vet. Rec.*, 165, 13-18.

Talling, J. C., Waran, N. K., Wathes, C. M. and Lines, J. A. (1996), 'Behavioural and physiological responses of pigs to sound', *Appl. Anim. Behav. Sci.*, 48, 187-202.

Tanida, H., Miura, A., Tanaka, T. and Yoshimoto, T. (1996), 'Behavioral responses of pig to darkness and shadows', *Appl. Anim. Behav. Sci.*, 49, 173-83.

Thodberg, K., Fogsgaard, K. K. and Herskin, M. S. (2019), 'Transportation of cull sows - deterioration of clinical condition from departure and until arrival at the slaughter plant', *Front. Vet. Sci.*, 6, 28 (open access). doi: 10.3389/fvets.2019.00028.

Torrey, S., Bergeron, R., Gonyou, H. W., Widowski, T. M., Lewis, N., et al. (2013a), 'Transportation of market-weight pigs: 1. Effect of season and truck type on behavior with a 2-hour transport', *J. Anim. Sci.*, 91, 2863-71.

Torrey, S., Bergeron, R., Faucitano, L., Widowski, T. M., Lewis, N., et al. (2013b), 'Transportation of market-weight pigs: 2. Effect of season and animal location in the truck on behavior with an 8-hour transport', *J. Anim. Sci.*, 91, 2872-8.

TQA (2016), *'National Pork Board Transport Quality Assurance Handbook'*, Version 6. http://www.pork.org/tqa-certification/tqa-program-materials/ (accessed on 13 May 2019).

Van de Perre, V., Permentier, L., de Bie, S., Verbecke, G. and Geers, R. (2010), 'Effect of unloading, lairage, pig handling, stunning and season on pH of pork', *Meat Sci.*, 86, 931-7.

Van der Wal, P. G., Engel, B. and Reimert, H. G. M. (1999), 'The effect of stress, applied immediately before stunning, on pork quality', *Meat Sci.*, 53, 101-6.

Van Putten, G. and Elshof, W. J. (1978), 'Observations on the effect of transportation on the well-being and lean quality of slaughter pigs', *Anim. Reg. Stud.*, 1, 247-71.

Vecerek, V., Malena, M., Malena Jr., M., Voslarova, E. and Chloupek, P. (2006), 'The impact of the transport distance and season on losses of fattened pigs during transport to the slaughterhouse in the Czech Republic in the period from 1997 to 2004', *Vet. Med.*, 51, 21-8.

Velarde, A. and Dalmau, A. (2012), 'Animal welfare assessment at slaughter in Europe: moving from inputs to outputs', *Meat Sci.*, 92, 244-51.

Published by Burleigh Dodds Science Publishing Limited, 2021.

Velarde, A., Gispert, M., Faucitano, L., Manteca, X. and Diestre, A. (2000), 'Survey of the effectiveness of stunning procedures used in Spanish pig abattoirs', *Vet. Rec.*, 146, 65-8.

Velarde, A., Gispert, M., Faucitano, L., Alonso, P., Manteca, X., et al. (2001), 'Effects of the stunning procedure and the halothane genotype on meat quality and incidence of haemorrhages in pigs', *Meat Sci.* 58, 313-19.

Velarde, A., Cruz, J., Gispert, M., Carrión, D., Ruiz-de-la-Torre, J. L., et al. (2007), 'Aversion to carbon dioxide stunning in pigs: effect of the carbon dioxide concentration and the halothane genotype', *Anim. Welf.*, 16, 513-22.

Vermeulen, L., Van de Perre, V., Permentier, L., De Bie, S., Verbeke, G., et al. (2015), 'Sound levels above 85 dB pre-slaughter influence pork quality', *Meat Sci.*, 100, 269-74.

Vitali, A., Lana, E., Amadori, M., Bernabucci, U., Nardone, A., et al. (2014), 'Analysis of factors associated with mortality of heavy slaughter pigs during transport and lairage', *J. Anim. Sci.*, 92, 5134-41.

Warriss, P. D. (1996), 'The consequences of fighting between mixed groups of unfamiliar pigs before slaughter', *Meat Focus Int.*, 5, 89-92.

Warriss, P. D. (2003), 'Optimal lairage times and conditions for slaughter pigs: a review', *Vet. Rec.*, 153, 170-6.

Warriss, P. D. and Brown, S. N. (1985), 'The physiological responses of fighting between in pigs and the consequences for meat quality', *J. Sci. Food Agric.*, 36, 87-92.

Warriss, P. D. and Brown, S. N. (1994), 'A survey of mortality in slaughter pigs during transport and lairage', *Vet. Rec.*, 134, 513-15.

Warriss, P. D. and Wilkins, L. J. (1987), 'Exsanguination in meat animals', *Seminar Preslaughter Stunning of Food Animals*, Brussels, Belgium, pp. 150-8.

Warriss, P. D., Brown, S. N., Bevis, E. A. and Kestin, S. C. (1990), 'The influence of pre-slaughter transport and lairage on meat quality in pigs of two genotypes', *Anim. Prod.*, 50, 165-72.

Warriss, P. D., Bevis, E. A., Edwards, J. E., Brown, S. N. and Knowles, T. G. (1991), 'Effect of the angle of slope on the ease with which pigs negotiate loading ramps', *Vet. Rec.*, 128, 419-21.

Warriss, P. D., Brown, S. N., Adams, S. J. M. and Corlett, I. K. (1994), 'Relationships between subjective and objective assessments of stress at slaughter and meat quality in pigs', *Meat Sci.*, 38, 329-40.

Weeks, C. A. (2008), 'A review of welfare in cattle, sheep, and pig lairages, with emphasis on stocking rates, ventilation and noise', *Anim. Welf.*, 17, 275-84.

Weeks, C. A., Brown, S. N., Warriss, P. D., Lane, S., Heasman, L., et al. (2009), 'Noise levels in lairages for cattle, sheep and pigs in abattoirs in England and Wales', *Vet. Rec.*, 165, 308-14.

Werner, C., Reiners, K. and Wicke, M. (2007), 'Short as well as long transport duration can affect the welfare of slaughter pigs', *Anim. Welf.*, 16, 385-9.

Weschenfelder, A. V., Torrey, S., Devillers, N., Crowe, T., Bassols, A., et al. (2012), 'Effects of trailer design on animal welfare parameters and carcass and meat quality of three Pietrain crosses being transported over a long distance', *J. Anim. Sci.*, 90, 3220-4676.

Weschenfelder, A. V., Torrey, S., Devillers, N., Crowe, T., Bassols, A., et al. (2013), 'Effects of trailer design on animal welfare parameters and carcass and meat quality of three Pietrain crosses being transported over a short distance', *Livest. Sci.*, 157, 234-44.

Wesoly, R., Jungbluth, I., Stefanski, V. and Weiler, U. (2015), 'Pre-slaughter conditions influence skatole and androstenone in adipose tissue of boars', *Meat Sci.*, 99, 6067.

WQ* (2009), 'Welfare Quality assessment protocol for pigs', Welfare Quality Consortium, Lelystad, the Netherlands, 122p.

Wotton, S. B. and Gregory, N. G. (1986), 'Pig slaughtering procedures: time to loss of brain responsiveness after exsanguination or cardiac arrest', *Res. Vet. Sci.*, 40, 148-51.

Xiong, Y, Green, A. and Gates, R. S. (2015), 'Characteristics of trailer thermal environment during commercial swine transport managed under U.S. industry guidelines', *Animals*, 5, 226-44.

Zhao, Y., Lu, X., Mo, D., Chen, Q. and Chen, Y. (2015), 'Analysis of reasons for sow culling and seasonal effects on reproductive disorders in southern China', *Anim. Rep. Sci.*, 159, 191-7.

Zurbrigg, K., van Dreumel, T., Rothschild, M. F., Alves, D., Friendship, R., et al. (2017), 'Pig-level risk factors for in-transit losses in swine: a review', *Can. J. Anim. Sci.*, 97, 339-46.

Chapter 4

Optimising the health of finisher pigs

Edgar Garcia Manzanilla, Pig Development Department, Teagasc, The Irish Agriculture and Food Development Authority, Ireland

1 Introduction

The growing-finishing period is the most expensive part of the pig production cycle accounting for 60-75% of the total costs depending on the type of production (Calderón Díaz et al., 2019). It should also be the easiest part to manage of the pig production cycle. Thus, it offers a big opportunity to maximise profit. Unfortunately, the performance of pigs in the finisher stage depends on the previous management. These aspects have been discussed in previous chapters. However, there are still a few factors that are important to optimise the health and productive performance of growing-finisher pigs. This chapter points out these factors and provides some insights on how to monitor and improve them.

2 How to measure pig health in grower-finisher phase

Health of pigs can be measured in different ways, and sometimes it is not measured properly but in the simplest way. Indeed, when there is an outbreak of an infectious disease, it is easy to detect and describe it. However, herds with

http://dx.doi.org/10.19103/AS.2022.0103.18

no outbreaks still have differences due to subclinical disease, stress or deficient management. Among the common variables used to measure health in finisher pigs, we could find feed intake, growth, feed conversion rate (FCR), time to slaughter, mortality, antimicrobial use (AMU) and lesions at slaughter (Agostini et al., 2014, 2015; O'Neill et al., 2021; Pessoa et al., 2021a,b; Rodrigues da Costa et al., 2020). Serology and pathogen detection are also very useful but are not always related to clinical disease. Indeed, all these variables can be affected by genetics, nutrition and climate, but once those are reasonably homogeneous, all the above variables can be used to compare health between herds and batches. The most important health measures for growing-finishing pigs are briefly described below:

1 Anorexia/hyporexia is one of the first signs of disease in animals. Pigs are no exception to this rule, and they tend to reduce activity, including feed intake, when they have any health problem and often before showing any other clinical signs. Thus, feed intake is a good way to measure how pigs are doing, although it should be combined with other indicators when possible.

2 Growth is indeed affected in poor health scenarios by the drop in intake, and this is also a health indicator to consider. Growth can also be affected by factors other than feed intake like severe activation of the immune system. However, in such cases, there would be other clinical signs that would be better to identify the health issue.

3 Feed efficiency, the relationship between intake and growth, is often assumed to be affected by the health status of the herd. However, in many cases, feed efficiency is more an indicator of issues with facilities and management (e.g. feeder type, feed presentation form, split sex and mixing) than of health. To some extent, it makes sense because hiporexia in sick animals is accompanied by a decrease in growth but does not necessarily affect FCR unless there is a severe disease. This might be counterintuitive for nutritionists since maintenance requirements are still constant and FCR should be affected (NRC, 2012). However, many field studies fail to identify the increase in FCR due to poor health status.

4 Time to slaughter is not a measure commonly included in scientific literature, but it is often an easier and more commercial way to detect issues in the growth of animals. Again, for the same genetics, diet and climate, herds with worse health status normally have problems selling pigs at the planned dates. This, at the same time, affects the farms and creates a lack of space, increases density and leaves less time for cleaning and disinfection. In current commercial pig farms, a period of 21 weeks to 23 weeks from birth to slaughter should be considered appropriate.

Pig farms with health problems are often around 25-27 weeks or even more weeks to reach the slaughter (Rodrigues da Costa et al., 2020).

5 Mortality and AMU are variables more directly related to clinical health issues and can be used for syndromic surveillance of herds (Lopes Antunes et al., 2017). Farms differ in their culling policy, and we need to consider mortality in all phases, including slaughterhouse, when looking at finisher mortality. The breakdown of the mortality causes is often not available, but it is very useful information to identify the aetiology of the problems, whether it is an infectious agent or problems in the facilities and management, and put in place the necessary corrective measures.

Other variables that are useful to measure the health status of finishing pigs and are easy to perform regularly are coughing measurements (Pessoa et al., 2021b; Fig. 1), lesions on farm or in slaughterhouse (van Staaveren et al., 2017) and, in the near future, farm biomarkers (López-Martínez et al., 2022).

3 Types of farms

The first thing to consider when planning how to manage our finisher stage to optimise health is the type of production structure. The management needed differs greatly between, for example, a multi-site production system and a farrow-to-finish farm where the finisher stage is metres away from the nursery (Fig. 2). In principle, no system is better than others if properly managed. In fact, literature favours one system or the other depending on the health indicator used (Casal et al., 2007; Fablet et al., 2018; Jager et al., 2012). For the chapter's purpose, the pros and cons of each system are presented.

Pigs in a farrow-to-finish farm are generally less stressed because no transport among production phases is required; they just walk to the next building. These pigs are also exposed to the pathogen burden to which they have already been exposed through their mothers, in the environment or equipment. On the other hand, a farrow-to-finish farm needs to be managed in a stricter way, trying to keep the breeding (gestation + lactation), the nursery and the finishers as separated units, so any outbreaks can be controlled easier. A multi-site system is ideal from an infectious disease management point of view if it is well run. It allows for easier all-in-all-out management and establishes a physical distance between the different phases. However, this advantage can be lost when animals from different origins are mixed or when the facility is managed with a continuous flow of animals. An inherent problem of the multi-site system is the stress caused by handling, transportation and relocation from one environment and management conditions to another, which can be a triggering factor for diseases promoted by stress such as Glässer's disease or ileitis.

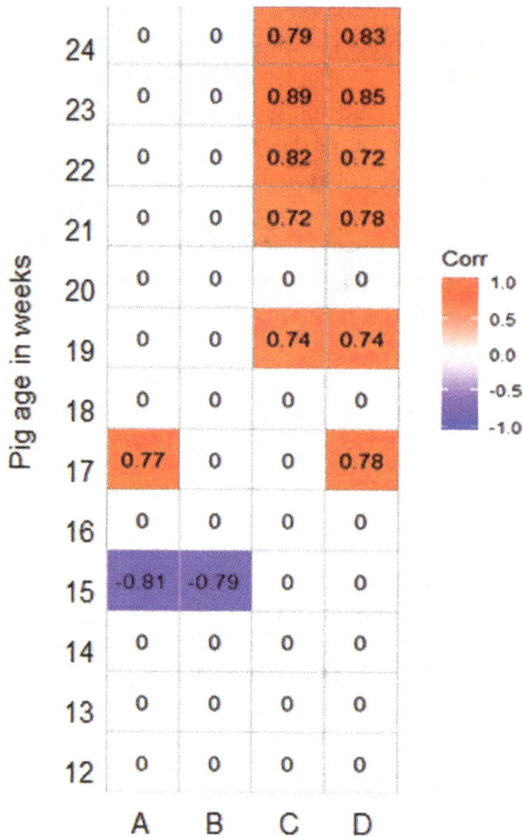

Figure 1 Lung lesions in slaughter can indicate disease as far as 8 weeks earlier in growing-finishing pigs. Automatic coughing monitors can detect disease on average 2 days before farm staff. In the figure, correlations between the prevalence of dorsocaudal (A) and cranial (B) pleurisy, scar lesions (C) and pneumonia (D) and the respiratory distress index measured automatically with coughing monitors along the growing-finishing stage are shown. Zeros correspond to non-significant results (Pessoa et al., 2021b).

4 Animal flow

In any type of farm, understanding how the animals flow through the system and segregating the different flows is one of the best ways to optimise health. Ideally, one batch of pigs should move all together (all-in-all-out), and those pigs brought to a hospital should never come back to the normal flow of pigs to avoid the spread of pathogens. The use of tags of different colours for each batch is probably the easiest way to keep batches separated and detect mixing. It is also easier to keep batches of pigs segregated and reduce disease incidence in farms working with batches at least 3 weeks apart from each other (Fablet et al., 2018). However, in many farms, the segregation between batches

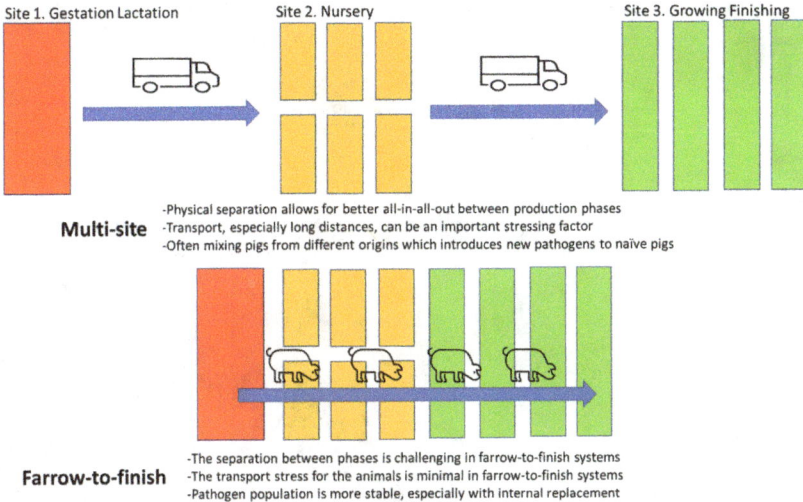

Figure 2 Pros and cons of the multi-site and farrow-to-finish pig systems.

is not respected. In Fig. 3, a real example of a farrow-to-finish farm is shown where one-weekly batch of 1096 pigs gets broken down into different flows as soon as the first week after weaning and along the production cycle (Rodrigues da Costa et al., 2021). In this example, pigs are divided at least into three main flows: pigs that move as planned, pigs that get delayed by 1 week and are mixed with the following batch and pigs that are delayed by several weeks and get mixed with several batches. Some of the pigs that are delayed by several weeks went through hospital rooms and went back to groups of healthy pigs. At slaughter, these delayed pigs were still in many cases sick and showed much higher levels of pleurisy, pericarditis, lameness and condemnations at slaughter than pigs that were not delayed (Calderón Díaz et al., 2017). Basically, the pathogens are being transmitted across different batches of naïve pigs. This may seem obvious to the veterinarian, but for a farm staff member with no training in infectious disease, this is the most normal thing to do. Farm staff need to be educated on the basics of infectious diseases and biosecurity.

Indeed, Fig. 3 is an extreme example where pigs are moved through four successive stages after weaning (nursery 1, nursery 2, growing and finishing). However, it serves the purpose of showing how wrong it can go and the need to avoid such practices. Probably the first thing to avoid is to design production cycles as complicated as this one. The more movements take place, the more stressed the pigs become and the more chances of mixing batches there are. Several epidemiological studies have shown strong associations between mixing growing-finishing pigs of different ages and higher incidence of diseases like enzootic pneumonia or pleurisy (Jager et al., 2012; Nathues et al., 2014).

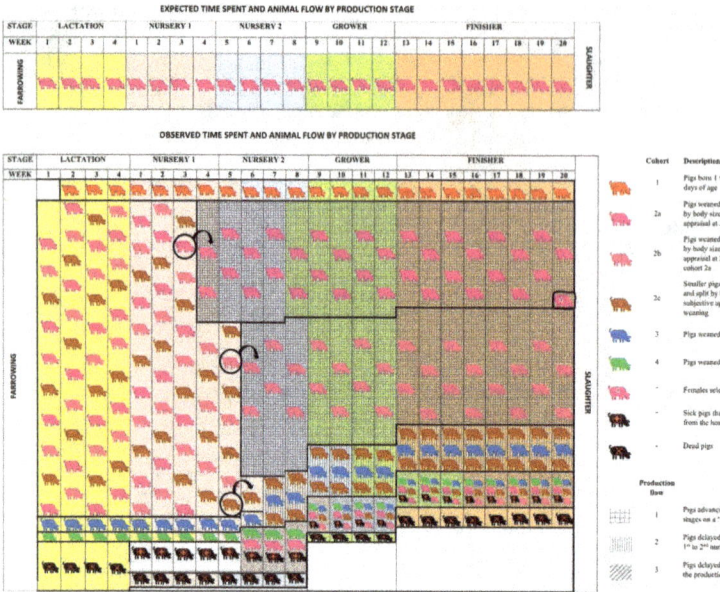

Figure 3 Expected vs. observed time spent and animal flow by production stage in a farrow-to-finish commercial pig farm where 1096 pigs were followed up from birth to slaughter to investigate management approaches for body weight variation and their possible implications for animal health (Rodrigues da Costa et al., 2021).

There are several causes of these different flows and mixing between ages. The two main causes are the natural variability of pig weights and the false impression that reorganising pigs by size helps reduce variability or improve performance (López-Vergé et al., 2018). Pigs' weights follow a normal distribution and, despite the positive correlation between birth and slaughter weight, reorganising pens by weight fails to reduce variability. In a recent unpublished study, 900 piglets were weighed individually and reorganised at weaning in 18 pens with average weights between 5 kg and 10 kg (weight range between pens: 5 kg) and weight ranges ((weight difference between the lower and higher weight in a pen)) within the pen of 300 g. By 16 weeks of age, the average weights of the pens ranged between 60 kg and 73 kg (13 kg), which is a 2.6-fold increase compared to weaning. However, the range of weights within pens was more than 20 kg, which is a 50-fold increase compared to weaning. The variability of weights at 16 weeks was the same in a control group of pigs not reorganised by weights. This proves the inefficacy of reorganising pigs to reduce variability. In fact, reorganising pigs by weight in the case of finishers has a negative effect, as proved by Camp Montoro et al. (2021, 2022). According to the results by Camp Montoro et al. (2021, 2022), mixing pigs in the growing-finishing stage has a more detrimental effect on growth than suboptimal space allowance and has an effect similar to using diets 10% deficient in lysine.

5 Stocking density

Stocking density or space allowance refers to the size of the area allocated to a pig and plays a critical role from a productive performance, animal health, welfare and economic standpoint (Thomas et al., 2017; Vermeer et al., 2017). Pig producers need to maximise production efficiency and minimise the housing cost, without compromising animal health. At the commercial scale, space allowance is adjusted by changing the number of pigs per pen. However, this method induces confounding on whether the pig is affected by space allowance, group size or feeder space. Flohr et al. (2018) stated that floor space is the main factor that influences grower-finisher pigs' performance; however, other factors can induce chronic stress by changes in the group structure and competition for resources. Kyriazakis and Whittemore (2006) came up with a formula where space allowance is expressed as:

$$\text{Space allowance}\left(m^2\right) = k \times BW^{0.667}$$

Table 1 Laying down minimum standards for the protection of nursery and grower-finisher pigs in the EU (Council of the European Union, 2008)

Body weight, kg	Space per pig, m²	k-value[a]	Space needed to keep k at 0.034[b]
Not more than 10	0.15	0.032	0.16
>10 but not more than 20	0.20	0.027	0.25
>20 but not more than 30	0.30	0.031	0.33
>30 but not more than 50	0.40	0.029	0.46
>50 but not more than 85	0.55	0.028	0.65
>85 but not more than 110	0.65	0.028	0.78
>110	1.00	0.039[c]	0.87

[a] $k - value = \dfrac{\text{Space allowance}}{\text{Body weight}^{0.667}}$; k-value was calculated based on the final body weight of the study.

[b] Based on Gonyou et al. (2006).

[c] k-value calculated based on pigs going to slaughter with a target body weight of 130 kg.

where k represents the space allowance coefficient and $BW^{0.667}$ means the geometric conversion of body weight (BW) in kilogram to area. The relationship between k-value and feed intake is highly similar for all pigs and, below a certain critical value, the pig reduces the feed intake proportionally, which is probably related to stress. Gonyou et al. (2006) reported a critical k-value of 0.0336; however, recent studies reported higher k critical values (Carpenter et al., 2018; Kim et al., 2017; Thomas et al., 2017). These values may depend on the type of facilities, environmental enrichment, pig genetics and, indeed, health status. It is likely that farms with lower health status are more affected by high stocking density because most of the research is done in high-health farms.

Directive 2008/120/EC establishes the minimum standards for pigs in the European Union (EU) in m²/pig (Table 1). For pig weights between 85 and 110 kg of BW, the minimum space is 0.65 m²/pig, which is a k-value = 0.028. Such legal minimum is clearly insufficient from a productive point of view and can also lead to adverse social behaviours that may affect pig health by chronic stress with clinical disease and higher condemnations at slaughter (Pandolfi et al., 2018). Table 1 shows a calculation of the minimum space recommended following Gonyou et al.'s (2006) critical k-value.

6 The importance of transfer weight and the transition diet

Agostini et al. (2014, 2015) studied the effects of different factors on the performance and mortality of 688 batches of pigs from nine different companies in Spain. The most relevant factor that affected mortality in all the companies was body weight at transfer from nursery to the growing phase, ranging from 15 to 25 kg. This is considered expectable because higher weights are probably related to a better ability of the pig to take the challenge of a new transport, new stressors and a new diet. Although research in this area is scarce, it is normal practice to not introduce the pig to the new diet directly but to try a feed transition during the first week in growing stage from the weaner diet to the new growing diet. The weights mentioned above (15–25 kg) are not the same in all the countries, and the later the transition between phases happens, the better for the pig from a health point of view.

Water, often referred to as the forgotten nutrient, becomes a key element in this stage. Dybkjær et al. (2006) reported that a low feed intake may be due to insufficient drinking activity, as solid feed intake must be accompanied by sufficient water intake. This is not as important for wet feed systems, although a separated drinker is always recommended. In practice, it is quite common not installing or restricting the flow of these drinkers to avoid water waste, especially when pigs play with the water. This type of action is detrimental to the pig, and the solution should be to provide other toys to the pigs or look for alternative drinker designs. The importance of dietary electrolyte balance has been mentioned in previous chapters for weaners and should also be considered here (Guzmán-Pino et al., 2015). Problems in water supply affect electrolyte balance and make pigs more sensitive to disease.

7 General control of infectious diseases

Previous chapters tackled different infectious diseases in detail, and the intention of this chapter is to describe the general aspects to minimise their occurrence in commercial farms. Randomised studies in these areas are difficult to perform, but there are quite a few epidemiological approaches that indicate some areas of importance. The main risk factors observed in epidemiological studies looking at digestive and respiratory infectious diseases in growing-finishing pigs are related to environmental control and biosecurity.

Ventilation is one of the areas that keeps appearing as a key aspect associated with health. It is well known that respiratory issues and mortality are increased during the colder quarters of the year (Agostini et al., 2015; Fablet et al., 2012). Thus, it is not surprising that automatic control of ventilation is associated with better health indicators in several studies (Agostini et al., 2014, 2015; Chantziaras et al., 2020; Matheson et al., 2022). Recommendations on temperature ranges and ventilation rates are included in Chapter 17. The technology for environmental control has improved significantly in recent years. The classic control of ventilation by temperature can now be easily improved by including sensors for humidity, CO_2 or NH_3 (Fig. 4). These changes can have a significant impact on the occurrence of infectious diseases. As an example, a high mean CO_2 concentration has been associated with a higher incidence of pneumonia (Fablet et al., 2012).

Another factor that is associated with bad health in epidemiological studies is the number of origins of the animals. Mixing piglets from different origins in any phase of production, but especially in growing-finishing pigs, is detrimental (Agostini et al., 2014, 2015; Fablet et al., 2018; Jager et al., 2012).

Cleaning and disinfecting of grower and finisher accommodation between groups is often considered less important than in previous stages of the production cycle, but it is related to better performance (van der Meer et al., 2016, 2017) and lower incidence of lesions like pleurisy (Jager et al., 2012).

Co-infections have been receiving more and more attention in recent years. Receiving pigs from weaner facilities infected with porcine reproductive and respiratory syndrome virus or porcine circovirus 2 is associated with lower performance and more pathology in the finisher stage (Fablet et al., 2018; Magalhaes et al., 2022). Management in such cases needs to be done considering the pathogens present on the farm. A correct application of

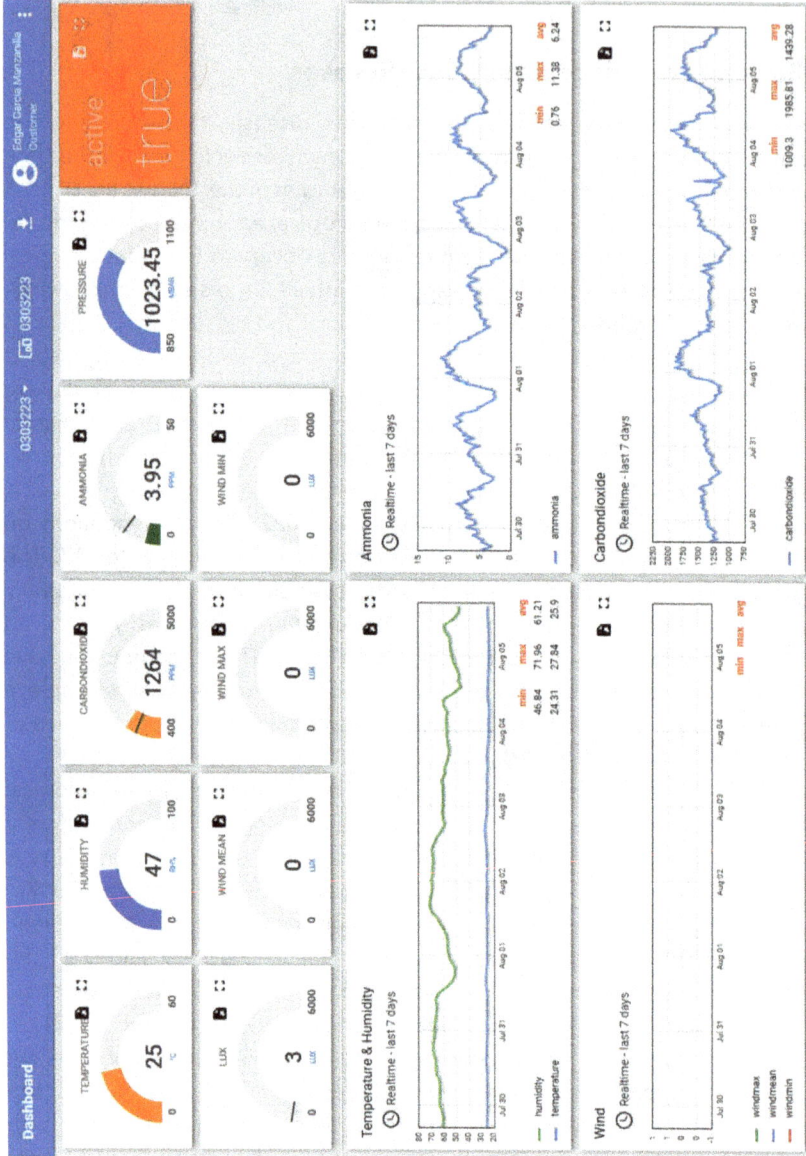

Figure 4 The current systems for environmental control allow the real-time monitoring of several parameters (e.g. temperature, humidity, NH_3, and CO_2) with access to any device and that can be used to automatically control the ventilation of the facilities.

biosecurity measures in these cases implies a knowledge of the epidemiology of the disease (Alarcón et al., 2021).

There are other aspects of management like the use of straw or other environmental enrichment (Matheson et al., 2022) or the herd size (Agostini et al., 2015), which are sometimes associated with the health status of the herd, but it is difficult to define whether they are positive or negative factors. In the case of herd size, there should not be any benefits related to the size if management is properly done. In the case of bedding and enrichment, in general, it should be beneficial. However, more studies are needed in this area.

8 Infectious vs. non-infectious diseases

Pig health management often focuses too much on infectious diseases. Probably, the main reason for this bias is the fact that more tools do exist to deal specifically with infectious diseases (antibiotics and vaccination, diagnostic analytical tests, etc.) than with non-infectious ones. However, growing-finishing pigs suffer from a series of diseases with no clear infectious component but with important prevalence and associated with high costs. Among these issues are stomach ulcers, hernias (inguinal or abdominal), prolapses, tail-biting lesions, lameness and arthritis (Heinonen et al., 2007; Teixeira et al., 2020). For many of these issues, there is not much literature available, and much more research is needed. The prevalence of most of these problems varies significantly among farms. For example, Gottardo et al. (2017) identified farm of origin as the main factor for gastric ulcers, with prevalence ranging from 0% to 36%. Robertson et al. (2002) also found the use of automated feeding systems, pelleted ration and changes in formulation and water from a dam rather than from a well or river as predisposing factors for stomach ulcers. Hernias, prolapses, lameness and arthritis are also typically problems observed in finishers more than in weaners, probably because of the size and weight of the animals. The heavy weight and fast growth of pigs in the grower-finisher stage exerts more physical pressure on these issues. Thus, whenever possible, these issues should be addressed at earlier stages to avoid their progress. Finally, a problem with no clear solution so far is tail biting (Taylor et al., 2010). Despite the extensive literature available in this area and the EU legislation ban on it, tail docking is so far the most effective measure to avoid tail biting in the current production systems. It is well known though that tail biting is associated with the occurrence of infectious diseases like abscessation in different parts of the body (Teixeira et al., 2016). Good health status is one of the preventive factors highlighted by Valros (2022), together with larger space allowance, partly slatted flooring, use of manipulable materials and use of long troughs for feeding. Obviously, all these improvements come with an extra cost that sometimes is difficult to justify (Niemi et al., 2021), and tail biting may still be a problem on some farms.

9 Conclusion

Based on the previous discussions, the general recommendations to optimise the health of pigs in growing-finishing stage include the following:

1 There is no good system or bad system. There is good management and bad management.
2 Bring the knowledge to the farm staff and make things easy for them. They take care of the details.
3 Define very well what health indicators you are using, why and how you will monitor them.
4 The health of growing-finishing pigs depends greatly on previous phases. Always keep in contact with those working with the breeding and the nursery farms and feedback information on the issues you are finding.
5 Put the animals to a good start bringing them into the growing stage with good weights and making a proper diet transition.
6 Avoid mixing pigs from different origins in the growing-finishing phase (or any phase).
7 Make the interval between batches as long as possible (ideally >3 weeks) and avoid mixing pigs of different ages to stop the transmission of infectious diseases between batches.
8 Avoid overstocking and always respect the ratio of pig:feeder space to reduce stress.
9 Always look for ways to improve environmental conditions for pigs and biosecurity. There is always room for improvement.

10 Where to look for further information

Grower-finisher pigs have been managed in the same way for more than 30 years. The development of new technologies and tools to manage their health will be key in the following years, especialy to anticipate health issues. However, research on new systems to raise pigs is needed to adapt pig farming to the needs of society and of an animal that, thanks to genetic selection, has little to do with what a pig was in the 90s.

 If you want to obtain further information or stay up to date on the most recent advances in this area, please follow the links included in this section and subscribe to their regular publications.

- https://extension.umn.edu/swine/swine-health.
- https://www.asi.k-state.edu/research-and-extension/swine/.
- https://www.ipic.iastate.edu/.
- https://eaphm.org/.
- https://www.teagasc.ie/animals/pigs/.

11 References

Agostini, P. S., Fahey, A. G., Manzanilla, E. G., O'Doherty, J. V., de Blas, C. and Gasa, J. 2014. Management factors affecting mortality, feed intake and feed conversion ratio of grow-finishing pigs. *Animal* 8(8):1312-1318.

Agostini, P. S., Manzanilla, E. G., de Blas, C., Fahey, A. G., da Silva, C. A. and Gasa, J. 2015. Managing variability in decision making in swine growing-finishing units. *Irish Veterinary Journal* 68(1):20.

Alarcón, L. V., Allepuz, A. and Mateu, E. 2021. Biosecurity in pig farms: a review. *Porcine Health Management* 7(1):5.

Calderón Díaz, J. A., Diana, A., Boyle, L. A., Leonard, F. C., McElroy, M., McGettrick, S., Moriarty, J. and Manzanilla, E. G. 2017. Delaying pigs from the normal production flow is associated with health problems and poorer performance. *Porcine Health Management* 3:13.

Calderón Díaz, J. A., Shalloo, L., Niemi, J. K., Kyriazakis, I., McKeon, M., McCutcheon, G., Bohan, A. and Manzanilla, E. G. 2019. Description, evaluation, and validation of the Teagasc Pig Production Model. *Journal of Animal Science* 97(7):2803-2821.

Camp Montoro, J., Boyle, L. A., Solà-Oriol, D., Muns, R., Gasa, J. and Manzanilla, E. G. 2021. Effect of space allowance and mixing on growth performance and body lesions of grower-finisher pigs in pens with a single wet-dry feeder. *Porcine Health Management* 7(1):7.

Camp Montoro, J., Pessoa, J., Solà-Oriol, D., Muns, R., Gasa, J. and Manzanilla, E. G. 2022. Effect of phase feeding, space allowance and mixing on productive performance of grower-finisher pigs. *Animals* 12(3):390.

Carpenter, C. B., Holder, C. J., Wu, F., Woodworth, J. C., Derouchey, J. M., Tokach, M. D., Goodband, R. D. and Dritz, S. S. 2018. Effects of increasing space allowance by removing a pig or gate adjustment on finishing pig growth performance. *Journal of Animal Science* 96(7):2659-2664.

Casal, J., Mateu, E., Mejia, W. and Martin, M. 2007. Factors associated with routine mass antimicrobial usage in fattening pig units in a high pig-density area. *Veterinary Research* 38(3):481-492.

Chantziaras, I., De Meyer, D., Vrielinck, L., Van Limbergen, T., Pineiro, C., Dewulf, J., Kyriazakis, I. and Maes, D. 2020. Environment-, health-, performance- and welfare-related parameters in pig barns with natural and mechanical ventilation. *Preventive Veterinary Medicine* 183:105150.

Council of the European Union (2008). COUNCIL DIRECTIVE 2008/120/EC of 18 December 2008. *Laying down minimum standards for the protection of pigs.* 47/5-47/13. Available at: https://eur-lex.europa.eu/legal-content/EN/TXT/PDF/?uri=CELEX:32008L0120&rid=19.

Dybkjær, L., Jacobsen, A. P., Tøgersen, F. A. and Poulsen, H. D. 2006. Eating and drinking activity of newly weaned piglets: effects of individual characteristics, social mixing, and addition of extra zinc to the feed. *Journal of Animal Science* 84(3):702-711.

Fablet, C., Dorenlor, V., Eono, F., Eveno, E., Jolly, J. P., Portier, F., Bidan, F., Madec, F. and Rose, N. 2012. Noninfectious factors associated with pneumonia and pleuritis in slaughtered pigs from 143 farrow-to-finish pig farms. *Preventive Veterinary Medicine* 104(3-4):271-280.

Fablet, C., Rose, N., Grasland, B., Robert, N., Lewandowski, E. and Gosselin, M. 2018. Factors associated with the growing-finishing performances of swine herds: an exploratory study on serological and herd level indicators. *Porcine Health Management* 4:6.

Flohr, J. R., Dritz, S. S., Tokach, M. D., Woodworth, J. C., Derouchey, J. M. and Goodband, R. D. 2018. Development of equations to predict the influence of floor space on average daily gain, average daily feed intake and gain:feed ratio of finishing pigs. *Animal* 12(5):1022-1029.

Gonyou, H. W., Brumm, M. C., Bush, E., Deen, J., Edwards, S. A., Fangman, T., McGlone, J. J., Meunier-Salaun, M., Morrison, R. B., Spoolder, H., Sundberg, P. L. and Johnson, A. K. 2006. Application of broken-line analysis to assess floor space requirements of nursery and grower-finisher pigs expressed on an allometric basis. *Journal of Animal Science* 84(1):229-235.

Gottardo, F., Scollo, A., Contiero, B., Bottacini, M., Mazzoni, C. and Edwards, S. A. 2017. Prevalence and risk factors for gastric ulceration in pigs slaughtered at 170 kg. *Animal* 11(11):2010-2018.

Guzmán-Pino, S. A., Solà-Oriol, D., Davin, R., Manzanilla, E. G. and Pérez, J. F. 2015. Influence of dietary electrolyte balance on feed preference and growth performance of postweaned piglets. *Journal of Animal Science* 93(6):2840-2848.

Heinonen, M., Hakala, S., Hämeenoja, P., Murro, A., Kokkonen, T., Levonen, K. and Peltoniemi, O. A. T. 2007. Case-control study of factors associated with arthritis detected at slaughter in pigs from 49 farms. *Veterinary Record* 160(17):573-578.

Jager, H. C., McKinley, T. J., Wood, J. L. N., Pearce, G. P., Williamson, S., Strugnell, B., Done, S., Habernoll, H., Palzer, A. and Tucker, A. W. 2012. Factors associated with pleurisy in pigs: a case-control analysis of slaughter pig data for England and Wales. *PLoS ONE* 7(2):e29655.

Kim, K. H., Kim, K. S., Kim, J. E., Kim, D. W., Seol, K. H., Lee, S. H., Chae, B. J. and Kim, Y. H. 2017. The effect of optimal space allowance on growth performance and physiological responses of pigs at different stages of growth. *Animal* 11(3):478-485.

Kyriazakis, I. and Whittemore, C. 2006. *Whittemore's Science and Practice of Pig Production* (3rd edn.). Blackwell Publishing Ltd., Oxford.

Lopes Antunes, A. C., Ersbøll, A. K., Bihrmann, K. and Toft, N. 2017. Mortality in Danish Swine herds: spatio-temporal clusters and risk factors. *Preventive Veterinary Medicine* 145:41-48.

López-Martínez, M. J., Franco-Martínez, L., Martínez-Subiela, S. and Cerón, J. J. 2022. Biomarkers of sepsis in pigs, horses and cattle: from acute phase proteins to procalcitonin. *Animal Health Research Reviews* 23(1):82-99.

López-Vergé, S., Gasa, J., Farré, M., Coma, J., Bonet, J. and Solà-Oriol, D. 2018. Potential risk factors related to pig body weight variability from birth to slaughter in commercial conditions. *Translational Animal Science* 2(4):383-395.

Magalhaes, E. S., Zimmerman, J. J., Thomas, P., Moura, C. A. A., Trevisan, G., Holtkamp, D. J., Wang, C., Rademacher, C., Silva, G. S. and Linhares, D. C. L. 2022. Whole-herd risk factors associated with wean-to-finish mortality under the conditions of a Midwestern USA swine production system. *Preventive Veterinary Medicine* 198:105545.

Matheson, S. M., Edwards, S. A. and Kyriazakis, I. 2022. Farm characteristics affecting antibiotic consumption in pig farms in England. *Porcine Health Management* 8(1):7.

Nathues, H., Chang, Y. M., Wieland, B., Rechter, G., Spergser, J., Rosengarten, R., Kreienbrock, L. and Grosse Beilage, E. 2014. Herd-level risk factors for the seropositivity to Mycoplasma hyopneumoniae and the occurrence of enzootic pneumonia Among fattening pigs in areas of endemic infection and high pig density. *Transboundary and Emerging Diseases* 61(4):316–328.

Niemi, J. K., Edwards, S. A., Papanastasiou, D. K., Piette, D., Stygar, A. H., Wallenbeck, A. and Valros, A. 2021. Cost-effectiveness analysis of seven measures to reduce tail biting lesions in fattening pigs. *Frontiers in Veterinary Science* 8:682330.

NRC 2012. *Nutrient Requirements of Swine* (11th edn.). National Academic Press, Washington, DC.

O'Neill, L., Calderón Díaz, J. A., Rodrigues da Costa, M., Oakes, S., Leonard, F. C. and Manzanilla, E. G. 2021. Risk factors for antimicrobial use on Irish pig farms. *Animals* 11(10):2828.

Pandolfi, F., Edwards, S. A., Maes, D. and Kyriazakis, I. 2018. Connecting different data sources to assess the interconnections between biosecurity, health, welfare, and performance in commercial pig farms in Great Britain. *Frontiers in Veterinary Science* 5:41.

Pessoa, J., McAloon, C., Rodrigues da Costa, M., Manzanilla, E. G., Norton, T. and Boyle, L. A. 2021a. Adding value to food chain information: using data on pig welfare and antimicrobial use on-farm to predict meat inspection outcomes. *Porcine Health Management* 7(1):55.

Pessoa, J., Rodrigues da Costa, M., Manzanilla, E. G., Norton, T., McAloon, C. and Boyle, L. A. 2021b. Managing respiratory disease in finisher pigs: combining quantitative assessments of clinical signs and the prevalence of lung lesions at slaughter. *Preventive Veterinary Medicine* 186:105208.

Robertson, I. D., Accioly, J. M., Moore, K. M., Driesen, S. J., Pethicka, D. W. and Hampson, D. J. 2002. Risk factors for gastric ulcers in Australian pigs at slaughter. *Preventive Veterinary Medicine* 53(4):293–303.

Rodrigues da Costa, M., Fitzgerald, R. M., Manzanilla, E. G., O'Shea, H., Moriarty, J., McElroy, M. C. and Leonard, F. C. 2020. A cross-sectional survey on respiratory disease in a cohort of Irish pig farms. *Irish Veterinary Journal* 73(1):24.

Rodrigues da Costa, M., Manzanilla, E. G., Diana, A., van Staaveren, N., Torres-Pitarch, A., Boyle, L. A. and Calderón Díaz, J. A. 2021. Identifying challenges to manage body weight variation in pig farms implementing all-in-all-out management practices and their possible implications for animal health: a case study. *Porcine Health Management* 7(1):10.

Taylor, N. R., Main, D. C. J., Mendl, M. and Edwards, S. A. 2010. Tail biting: a new perspective. *Veterinary Journal* 186(2):137–147.

Teixeira, D. L., Harley, S., Hanlon, A., O'Connell, N. E., More, S. J., Manzanilla, E. G. and Boyle, L. A. 2016. Study on the association between tail lesion score, cold carcass weight, and viscera condemnations in slaughter pigs. *Frontiers in Veterinary Sciences* 3:24.

Teixeira, D. L., Salazar, L. C., Enriquez-Hidalgo, D. and Boyle, L. A. 2020. Assessment of animal-based pig welfare outcomes on farm and at the abattoir: a case study. *Frontiers in Veterinary Science* 7:576942.

Thomas, L. L., Goodband, R. D., Woodworth, J. C., Tokach, M. D., Derouchey, J. M. and Dritz, S. S. 2017. Effects of space allocation on finishing pig growth performance and carcass characteristics. *Translational Animal Science* 1(3):351–357.

Valros, A. 2022. Review: the tale of the Finnish pig tail – how to manage non-docked pigs? *Animal* 16 (Suppl. 2):100353.

van der Meer, Y., Gerrits, W. J. J., Jansman, A. J. M., Kemp, B. and Bolhuis, J. E. 2017. A link between damaging behaviour in pigs, sanitary conditions, and dietary protein and amino acid supply. *PLoS ONE* 12(5):e0174688.

van der Meer, Y., Lammers, A., Jansman, A. J. M., Rijnen, M. M. J. A., Hendriks, W. H. and Gerrits, W. J. J. 2016. Performance of pigs kept under different sanitary conditions affected by protein intake and amino acid supplementation. *Journal of Animal Science* 94(11):4704–4719.

van Staaveren, N., Doyle, B., Manzanilla, E. G., Calderón Díaz, J. A., Hanlon, A. and Boyle, L. A. 2017. Validation of carcass lesions as indicators for on-farm health and welfare of pigs. *Journal of Animal Science* 95(4):1528–1536.

Vermeer, H. M., Dirx-Kuijken, N. C. P. M. M. and Bracke, M. B. M. 2017. Exploration feeding and higher space allocation improve welfare of growing-finishing pigs. *Animals* 7(5):3–11.

www.ingramcontent.com/pod-product-compliance
Lightning Source LLC
Chambersburg PA
CBHW050536270326
41926CB00015B/3252